Periodic Table of the Elements with the Gmelin System Numbers

1 H 2																	2 He 1
3 Li 20	4 Be 26											5 B 13	6 C 14	7 N 4	8 O 3	9 F 5	10 Ne 1
11 Na 21	12 Mg 27											13 Al 35	14 Si 15	15 P 16	16 S 9	17 Cl 6	18 Ar 1
19 * K 22	20 Ca 28	21 Sc 39	22 Ti 41	23 V 48	24 Cr 52	25 Mn 56	26 Fe 59	27 Co 58	28 Ni 57	29 Cu 60	30 Zn 32	31 Ga 36	32 Ge 45	33 As 17	34 Se 10	35 Br 7	36 Kr 1
37 Rb 24	38 Sr 29	39 Y 39	40 Zr 42	41 Nb 49	42 Mo 53	43 Tc 69	44 Ru 63	45 Rh 64	46 Pd 65	47 Ag 61	48 Cd 33	49 In 37	50 Sn 46	51 Sb 18	52 Te 11	53 I 8	54 Xe 1
55 Cs 25	56 Ba 30	57** La 39	72 Hf 43	73 Ta 50	74 W 54	75 Re 70	76 Os 66	77 Ir 67	78 Pt 68	79 Au 62	80 Hg 34	81 Tl 38	82 Pb 47	83 Bi 19	84 Po 12	85 At 8a	86 Rn 1
87 Fr 25a	88 Ra 31	89*** Ac 40	104 71	105 71													

*Lanthanides 39

58 Ce	59 Pr	60 Nd	61 Pm	62 Sm	63 Eu	64 Gd	65 Tb	66 Dy	67 Ho	68 Er	69 Tm	70 Yb	71 Lu

***Actinides

90 Th 44	91 Pa 51	92 U 55	93 Np 71	94 Pu 71	95 Am 71	96 Cm 71	97 Bk 71	98 Cf 71	99 Es 71	100 Fm 71	101 Md 71	102 No 71	103 Lr 71

* NH₄ 23 → NH_4 23

A Key to the Gmelin System is given on the Inside Back Cover

Gmelin Handbook of Inorganic Chemistry

8th Edition

Gmelin Handbook
of Inorganic Chemistry

8th Edition

Gmelin Handbuch der Anorganischen Chemie

Achte, völlig neu bearbeitete Auflage

Prepared
and issued by

Gmelin-Institut für Anorganische Chemie
der Max-Planck-Gesellschaft
zur Förderung der Wissenschaften

Director: Ekkehard Fluck

Founded by — Leopold Gmelin

8th Edition — 8th Edition begun under the auspices of the
Deutsche Chemische Gesellschaft by R. J. Meyer

Continued by — E. H. E. Pietsch and A. Kotowski, and by
Margot Becke-Goehring

Springer-Verlag Berlin Heidelberg GmbH 1987

Gmelin Handbook of Inorganic Chemistry

8th Edition

Th
Thorium

Supplement Volume C 3

Compounds with Nitrogen

With 38 illustrations

AUTHORS

R. Benz, A. Naoumidis
Kernforschungsanlage Jülich
Jülich, Federal Republic of Germany

D. Brown
Chemistry Division, Atomic Energy Research Establishment
Harwell, England

CHIEF EDITORS

Rudolf Keim, Gmelin-Institut, Frankfurt am Main

Cornelius Keller, Supervising Scientific Coordinator
for the Thorium Supplement Volumes,
Schule für Kerntechnik, Kernforschungszentrum Karlsruhe

System Number 44

Springer-Verlag Berlin Heidelberg GmbH 1987

LITERATURE CLOSING DATE: MID OF 1986
IN SOME CASES MORE RECENT DATA HAVE BEEN CONSIDERED

ISBN 978-3-662-06332-3   ISBN 978-3-662-06330-9 (eBook)
DOI 10.1007/978-3-662-06330-9

Library of Congress Catalog Card Number: Agr 25-1383

© by Springer-Verlag Berlin Heidelberg 1987
Originally published by Springer-Verlag Berlin Heidelberg New York in 1987
Softcover reprint of the hardcover 8th edition 1987

Preface

This volume C 3, as a part of the Gmelin "Thorium" Handbook, Series C, describes the thorium-nitrogen compounds. Included are compounds both of technological importance like the nitrides and the nitrates and those of merely scientific interest, such as amides and related compounds. However, due to the decreasing technical importance of the nuclear thorium fuel cycle, especially with the advanced fuels like the nitride ThN, in recent years, the thorium compounds with nitrogen have been investigated much less extensively than the corresponding uranium compounds. In order to have the data for the Th–N–X systems accumulated in one specific volume, the decision was made to publish this volume without incorporating other Th systems.

ThN is the compound with the lowest N:Th ratio. In addition to its (former) nuclear interest due to its thermal and radiation stability, it has many very interesting physicochemical properties. Thorium nitrate, the other well-investigated compound, is of importance because it is (in the form of an adduct with tri-n-butylphosphate) the extracted compound when burnt-up thorium fuels are reprocessed.

Despite the wealth of accumulated data on the chemical and physicochemical properties of the compounds discussed, the knowledge of the compounds and of the systems is far from satisfactory – it must be deepened and improved in further studies.

I would like to thank the competent authors for their critical contributions as well as the Gmelin-Institute for the excellent cooperation provided, especially Prof. Dr. Fluck and Dr. Keim, the editor-in-chief of this volume.

Karlsruhe
June 1987   Cornelius Keller

Volumes published on "Radium and Actinides"

Table of Contents

4 Compounds of Thorium with Nitrogen

4.1 Binary Nitrides

Benz, R., Naoumidis, A.
Kernforschungsanlage Jülich
Jülich, Federal Republic of Germany

4.1.1 Overview and General Remarks

The chemical properties of thorium differ from those of the other light members of the actinide series in exhibiting the almost unique valence 4+ (exclusively 4+ in the nitrides at sufficiently high N_2 pressures) with the exceptions of ThN, ThS, and ThI_2. Chemical bonds formed by Th show very little f character, the 5f atomic orbitals forming only high-energy delocalized bonds. With increasing atomic number in the first half of the actinide series, there is a systematic variation in the stability of other ion valences and in multiplicity of valences. Simultaneously, there is an increasing participation of 5f orbital hybridization in bonds [1].

The properties of the thorium nitrides can be compared to those of the much more investigated ones of uranium as reviewed in the Gmelin Handbook [2]. The two mononitrides melt at almost the same temperature and N_2 pressure conditions, 2820°C under 2.6 atm N_2 for ThN as compared to 2830°C under 3.5 atm N_2, suggesting very similar high-temperature thermodynamic stabilities. Whereas ThN has no localized paramagnetism, the free U atom with $5f^36d7s^2$ occupied orbitals does have 5f electrons which can participate in chemical bonding, and UN shows a weak ferromagnetic ordering [3]. All the valence electrons of Th have nearly the same energy [4] and are readily released to form stoichiometric Th_3N_4, stable over a range of N_2 pressures. Uranium releases at most 88% of its valence electrons to form the UN_{2-x} phase of variable composition, and except possibly at elevated N_2 pressures, never reaches the UN_2 composition corresponding to the maximum possible valence of U.

The lanthanide elements also form mononitrides and with increasing atomic number the electron occupation of f orbitals increases, but the electron structure is distinctly different from that of the actinides. The electrons in occupied f orbitals of the lanthanides are largely responsible for the magnetic properties but the radial extent of the f orbitals is less localized and the energy states are so low that the electrons participate very little in chemical bonding. The free Ce atom, the analog of Th, is an exception. It has 4 outer-shell electrons in the $4f^15d^16s^2$ orbitals. The 4f state is only slightly higher in energy than the 5d state and Ce can exhibit either the 3+, e.g. Ce_2O_3, or 4+ valence state, e.g. CeO_2 and CeF_4. X-ray photoemission data show, however, that the Ce 4f levels remain localized in both Ce metal and in CeN [5]. No binary nitride higher than the mononitride has been reported. However, Ce does appear to have the 4+ valence state in the reported ternary compound Li_2CeN_2 [6, 7].

Many of the Th nitrides are more or less stable at elevated temperatures, electrically and thermally insulating, and have an extremely weak (Van Vleck) temperature-independent paramagnetism. The face-centered cubic (fcc) mononitride, ThN, is an important exception. It is metallic. It has a weak temperature-independent (Pauli) paramagnetism. Subsequent comparison of the electronic properties will show many simularities between ThN and metallic Th. An appreciable metallic bonding is also indicated by anomalously large interatomic distances in ThN.

Because of their high density and good thermal conductivity, ThN alloys have been considered as possible breeder material for high-temperature reactor technology [8, 9]. On reacting with a neutron, Th is converted by the nuclear reaction $^{232}Th(n,\gamma)^{233}Th \xrightarrow{\beta^-} {}^{233}Pa \xrightarrow{\beta^-} {}^{233}U$ to the fissionable nuclide ^{233}U capable of sustaining a thermal nuclear reaction. It does not occur in very rich deposits, but the estimated average concentration of Th in the earth's igneous rocks is more than 3 times that of U [10 to 12], making Th potentially capable of supplementing the earth's future supply of fissionable material. To date, however, only the oxides and carbides of Th have been used in experimental breeder reactors. Although the nitrides of Th have the drawback of being sensitive to moisture and O_2 in air, they have favorable thermal transport properties that could be economically useful. As yet, however, there is no major technological application, but this situation could change in the future if the world's energy needs become sufficiently demanding.

Previous reviews of thorium nitrides are the early work in the Gmelin Handbook [13], the physical properties as tabulated by Peterson, Curtis [14], the tabulation and contributions to the thermochemical data by Rand [15], their brief mention in an engineering data tabulation of nitrides [16], and a chapter on thorium nitrides in "High-Temperature Nuclear Fuel" [17].

References for 4.1.1:

[1] Brooks, M. S. S. (J. Magn. Magn. Mater. **29** [1982] 257/61; J. Phys. F **14** [1984] 857/71).
[2] Gmelin Handbook "Uranium" Suppl. Vol. C7, 1981, pp. 1/93.
[3] Lam, D. J., Aldred, A. T. (in: Freemann, A. J., Darby, J. B., The Actinides: Electronic Structure and Related Properties, Magnetic Properties of Actinide Compounds, Academic, New York 1974, pp 109/79).
[4] Keeton, S. C., Loucks, T. L. (Phys. Rev. [2] **146** [1966] 429/31).
[5] Baer, Y., Hauger, H., Zürcher, Ch., Campagna, M., Wertheim, G. K. (Phys. Rev. [3] B **18** [1978] 4433/9).
[6] Barker, M. G., Alexander, I. C. (J. Chem. Soc. Dalton Trans **1974** 2166/70).
[7] Halot, D., Flahaut, J. (Compt. Rend. C **272** [1971] 465/7).
[8] Bauer, A. A., Moak, D. P., Alexander, C. A. (Trans. Am. Nucl. Soc. **27** [1977] 287).
[9] Trauger, D. B. (Ann. Nucl. Energy **5** [1978] 375/403).
[10] Grayson, M., Eckroth, D. (Kirk-Othmer Encycl. Chem. Technol. 3rd Ed. **22** [1983] 990).

[11] Gmelin Handbuch "Thorium" 1955, pp. 5/32.
[12] Rankama, K., Sahama, Th. G. (Geochemistry, Univ. Chicago Press, Chicago 1955, pp. 570/1).
[13] Gmelin Handbuch "Thorium" 1955, pp. 241/3.
[14] Peterson, S., Curtis, C. E. (ORNL-4503 [1973] Vol. 2, pp. 1/20; C.A. **80** [1974] No. 123945).
[15] Rand, M. H. (At. Energy Rev. Spec. Issue No. 5 [1975] 7/85).
[16] Battelle Columbus Labs. (MCIC-HB-07 – Vol. 1 [1976] 1/119; INIS Atomindex **7** [1976] No. 265 237).

[17] Kotel'nikov, R. B., Bashlykov, S. N., Kashtanov, A. I., Men'shikova, T. S. (High-Temperature Nuclear Fuel, 2nd Ed., Chapter 10, [Vysokotemperaturnoe Yadernoe Toplivo] Atomizdat, Moscow 1978, pp. 370/4).

4.1.2 The Thorium-Nitrogen System

4.1.2.1 Electron Transfer Between Gaseous Nitrogen and Thorium

Cross sections for charge transfer from N^+, N_2^+, O^+, and CO_2^+ ions to gaseous Th and U atoms were determined as a function of the ion energy, 1 to 500 eV, by a crossed ion modulated neutral beam method. The cross-section values were all found to increase monotonically with decreasing ion energy with the following values as read from plots: 3 to 10 ($N^+ + Th = N + Th^+$) and 0.5 to 2 ($N_2^+ + Th = N_2 + Th^+$). This dependence on the projectile ion energy indicates that the excess energy is absorbed by internal excitation of the product particles, especially by excitation of the low-lying states of the Th ions.

Reference for 4.1.2.1:

Rutherford, J. A., Vroom, D. A. (J. Chem. Phys. **69** [1978] 332/5).

4.1.2.2 Chemical Diffusion of Nitrogen in Thorium

α-Th. The volume rates of N_2 uptake when iodide-crystal-bar Th metal is heated in 1 atm N_2 at various temperatures from 670 to 1490°C have been reported by Gerds, Mallett [1]. The volume uptake, V_{N_2}, varies parabolically with time as $V_{N_2}^2 = kt$, up to a critical time, after which the dependence is linear. The parabolic rate constants describing the N_2 uptake as listed in Table 1 are represented by the equation k (in $(mL \cdot cm^{-2})^2 \cdot s^{-1}$) $= 5.9 \cdot exp(-24300/RT)$ in the temperature range 943 to 1763 K, where the energy of activation is given in cal/mol N_2.

The composition-average diffusion coefficients of nitrogen in α-Th, $\overline{D}_N$(α-Th), were calculated from analyses of the nitrogen distribution in the Th metal. The values as illustrated in **Fig. 1** are represented by the equation $\overline{D}_N$(α-Th) (in cm^2/s) $= 2.1 \times 10^{-3} \cdot exp(-22500/RT)$; temperature range 1118 to 1763 K, where the energy of activation is $E_A = 22500$ cal/g-atom N. The diffusion of N atoms in the α-Th lattice is seen in Fig. 1 to be some 10^3 times faster than in the ThN lattice having the same metal atom configuration [1].

Table 1

Parabolic Rate Constants, k, Describing the Rates of Reaction of N_2 with Different Samples of Th Metal [1].

temperature in °C	670	730	900	900	960	1020	1100
k in 10^{-4} $(mL/cm^2)^2/s$	0.13	0.28	2.94	4.00	1.97	3.55	6.00

temperature in °C	1175	1200	1200	1255	1370	1490
k in 10^{-4} $(mL/cm^2)^2/s$	11.9	6.62	9.83	22.3	36.0	130

β-Th. Rates of N diffusion in β-Th, $\overline{D}_N$(β-Th), have been calculated from measurements of N electrotransport rates in β-Th and reported by Peterson et al. [2] and Peterson, Carnahan [3], and reviewed by Smith et al. [4]. The latter results are represented by the equation $\overline{D}_N$(β-Th) (in cm²/s) = $3.2 \times 10^{-3} \cdot \exp(-17000/RT)$ in the temperature range 1713 to 1988 K, where the activation energy is 17000 cal/g-atom N [3].

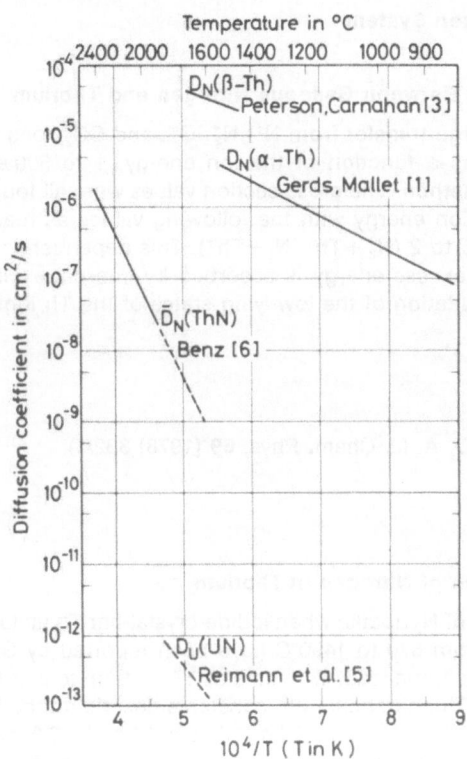

Fig. 1

Average chemical diffusion coefficients of N in α-Th [1], β-Th [3], and in ThN [6]. The diffusion coefficient of U in UN is given for comparison.

References for 4.1.2.2:

[1] Gerds, A. F., Mallett, M. W. (J. Electrochem. Soc. **101** [1954] 175/80).
[2] Peterson, D. T., Schmidt, F. A., Verhoeven, T. D. (Trans AIME **236** [1966] 1311/5).
[3] Peterson, D. T., Carnahan, T. (Trans. Met. Soc. AIME **245** [1969] 213/5).
[4] Smith, J. F., Carlson, O. N., Peterson, D. T., Scott, T. E. (Thorium: Preparation and Properties, Chapter 5, Iowa State Univ. Press, Ames, Ia., 1975, pp. 175/86).
[5] Reimann, D. K., Kroeger, D. M., Lundy, T. S. (J. Nucl. Mater. **38** [1971] 191/6).
[6] Benz, R. (J. Electrochem. Soc. **119** [1972] 1596/602).

4.1.2.3 Phase Relations in the Thorium-Nitrogen System

The binary Th–N phase diagram is illustrated in **Fig. 2**. Two equilibrium compounds are indicated, ThN [1] with a NaCl-type crystal structure [2] and Th_3N_4 [3] recently designated as α-Th_3N_4 [4] with a trigonal structure [5].

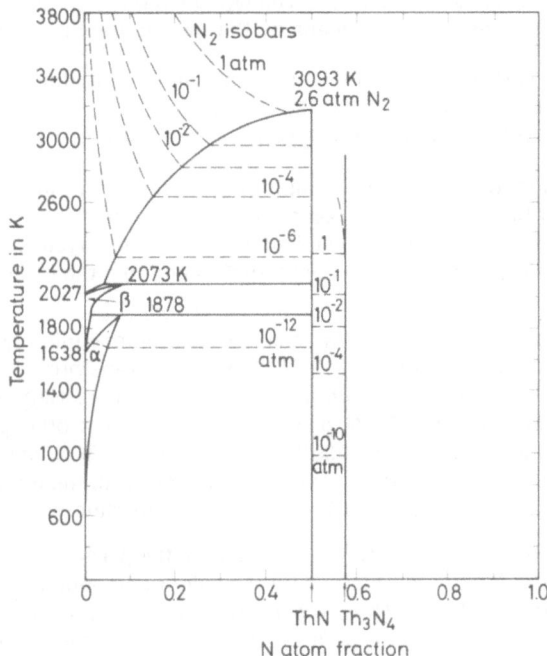

Fig. 2

The Th–N phase diagram, drawn by the authors of this article.
The various sources of data are as follows:

the Th(l)–ThN two-phase region data are
  a) the 2.6-atm isobar at the ThN melting point [9];
  b) the 10^{-4} to 1-atm isobars [12];
  c) the Gibbs free energy of formation of ThN,
     $\Delta G_f = -88\,000 + 20.24\,T$ cal/mol (cf. p. 21);

the ThN–Th_3N_4 phase region data were calculated with the equation
$\Delta G_f\,(Th_3N_4) = -308\,000 + 80.58\,T$ cal/mol, cf. p. 40.

Th–ThN Region. Nitrogen dissolves in metallic Th raising the temperature of the allotropic transformation, α-Th $\rightarrow$ β-Th, from 1365 to $1605 \pm 20°C$ where the solubility in α-Th is $7.4 \pm 0.2\%$ N [6]. The temperature dependence of the N solubility in α-Th has been measured by Gerds, Mallett [7] and is given by the equation

$$\log c_o = -2405/T + 0.9115 \quad (T = 1118 \text{ to } 1763 \text{ K}),$$

where c_o denotes the concentration of nitrogen in wt%. From this equation, the molar heat of solution is $\Delta H_s = +11\,000 \pm 400$ cal/mol N_2.

References for 4.1.2.3 see p. 6

The melting point of pure β-Th metal, $1754 \pm 15°C$, is increased by N additions [1, 8] to a peritectic point at $1800 \pm 25°C$ [6]. The peritectic reaction on cooling is

$$Th(l) + ThN \rightarrow \beta\text{-}Th(+N)$$

where the N solubility in β-Th is 8.3 ± 0.2 at%. The liquidus above the peritectic temperature has been measured with quenched specimens by the metallographic technique [6].

ThN Phase. The fcc ThN phase is a line compound with the limiting N/Th ratios fixed at $\geqq 0.975 \pm 0.01$ [9, p. 98] and $\leqq 1.000 \pm 0.005$ at 2000°C [11, footnote on p. 424]. The ThN phase melts congruently at $2820 \pm 35°C$ in 2.6 atm N_2 with the composition $ThN_{0.995 \pm 0.005}$ [6, 9, 12].

Th_3N_4 Phase. The Th_3N_4 phase is a line compound with the N/Th ratio of 1.33 ± 0.02 up to about 1900°C [3, 6, 10]. Above 1900°C, this compound becomes N-deficient as is indicated by a very coarse Widmannstätten structure exhibited by quenched crystals [6]. The structure suggests that the N atoms are very mobile above 1900°C and that the N-deficient phase does not quench in.

Metastable β-Th_3N_4. A metastable β-Th_3N_4 phase with a monoclinic crystal structure has been reported by Juza, Gerke [4]. Above 1120°C, this phase transforms irreversibly into the equilibrium Th_3N_4 phase which these authors denote as α-Th_3N_4. High-temperature X-ray diffraction traces have shown that no Th_3N_4 polymorphic form and no other phase between the ThN and Th_3N_4 compositions exist up to 1905°C [13]. Thus, the β-Th_3N_4 compound cannot be assigned to the equilibrium phase diagram which is in accord with the reported metastability [4]. The designation α for the equilibrium Th_3N_4 phase is superfluous.

Pialoux [14] reports having confirmed the existence of the β-Th_3N_4 compound whereby it was identified by high-temperature diffractometry as a minor component in some C-saturated hexagonal "β-ThCN" + rhombohedral Th_3N_4 specimens at temperatures of 1300 to 1500°C. The β-Th_3N_4 was never formed directly from "β-ThCN" but could be formed starting with rhombohedral Th_3N_4 (his α-Th_3N_4). In addition, the β-Th_3N_4 phase was confirmed to persist in 1 atm N_2 at 1450°C after 19 h, during which time the coexisting "β-ThCN" transformed to Th_3N_4 (α-Th_3N_4). Pialoux [14] regards this as support for its metastability but the possibility that it is some ternary compound does not seem to be excluded.

References for 4.1.2.3:

[1] Chiotti, P. (AECD-3072 [1951] 1/63; N.S.A. **5** [1951] No. 3141).
[2] Rundle, R. E. (Acta Cryst. **1** [1948] 180/7).
[3] Matignon, C., Delépine, M. (Ann. Chim. Phys. [8] **10** [1907] 130/44).
[4] Juza, R., Gerke, H. (Z. Anorg. Allgem. Chem. **363** [1968] 245/57).
[5] Benz, R., Zachariasen, W. H. (Acta Cryst. **21** [1966] 838/40).
[6] Benz, R., Hoffmann, C. G., Rupert, G. N. (J. Am. Chem. Soc. **89** [1967] 191/7).
[7] Gerds, A. F., Mallett, M. W. (J. Electrochem. Soc. **101** [1954] 175/80).
[8] Wilhelm, H. A., Chiotti, P. (Trans. Am. Soc. Metals **42** [1950] 1295/310).
[9] Benz, R. (J. Nucl. Mater. **31** [1969] 93/8, see addendum to [6] therein).
[10] Aronson, S., Auskern, A. B. (J. Phys. Chem. **70** [1966] 3937/41).

[11] Benz, R., Troxel, J. E. (High Temp. Sci. **3** [1971] 422/32).
[12] Olson, W. M., Mulford, R. N. R. (J. Phys. Chem. **69** [1965] 1223/6).
[13] Benz, R., Balog, G. (High Temp. Sci. **3** [1971] 511/22).
[14] Pialoux, A. (J. Nucl. Mater. **91** [1980] 127/46).

4.1.2.4 Changes in Properties of Thorium by Nitrogen

Crystal Structure

Both of the allotropic forms of Th can dissolve an appreciable amount of N. Incorporation of N in crystals of the low-temperature α-Th form (fcc, Cu-type crystal structure, space group Fm3m (No. 225)) increases the lattice spacing, but metallographic examination shows that most of the dissolved N precipitates out as ThN on cooling at ordinary cooling rates.

The α-Th in 1-gram Th + ThN specimens after equilibrating at elevated temperatures and quenching in a stream of He gives broad X-ray reflections, and the room-temperature lattice parameter shows considerable scatter, 509.0 ± 0.9 pm [1]. These values tend to be higher than those reported for van Arkel-deBoer Th, 508.4 ± 0.2 [2], 508.43 ± 0.01 [3], 508.40 ± 0.01 [4], and 508.5 ± 0.1 pm [5]. A more systematic dependence of a on N content has been reported by McLachlan [6] in an investigation of alloys with low N content of 0.04, 0.16, 0.32, 0.58, and 0.97 at% N. The lattice parameter of these alloys after homogenization at 1000°C and water quenching shows a linear increase with N content up to 508.8 pm at 1 at% N. The room-temperature results are represented by the equation

$$a \text{ (in pm)} = 508.45 + 34 \cdot x_N$$

in the range of N atom fractions $x_N = 0.0004$ to 0.0097. Alloys that are equilibrated at 900°C and furnace-cooled follow this same line only up to $x_N = 0.0038$ and form a plateau with a = 508.58 pm at higher x_N values. The N solubility limit is thus $x_N = 0.0038$ at 900°C [6].

At elevated temperatures, above 900°C, presence of dissolved N increases the α-Th lattice spacing over that of the pure metal to a = 521.4 pm at 1605°C. Whereas the pure metal transforms $\alpha \to \beta$ with a volume increase of +0.16 cm³/mol (1365 ± 15°C) [5, 7], the ThN-saturated α-Th transforms to the β-allotrope having very little dissolved N with a volume decrease of −0.22 cm³/mol at 1605 ± 20°C (a of N-containing β-Th at 1605°C is 412.4 pm). The dissolution of N in the α-Th lattice is an endothermic reaction that leads to a stronger binding between the Th atoms [7].

Lattice Thermal Expansion

High-temperature lattice parameters of N-saturated α-Th are nearly the same as those of pure Th metal up to nearly 1000°C [7]. The reported data for pure α-Th as a function of temperature can be represented to within 1% by the following equations:

$a \text{ (in pm)} = 508.307 + 3.6778 \times 10^{-3}\, t + 25.236 \times 10^{-6}\, t^2 + 49.4825 \times 10^{-9}\, t^3$ (t = 10 to 60°C) [3];

$a \text{ (in pm)} = 508.51 + 3.89 \times 10^{-3}\, t + 6.28 \times 10^{-6}\, t^2$ (t = 25 to 419°C), calculated by the authors of this article from the data in [8];

$a \text{ (in pm)} = 508.3 + 6.040 \times 10^{-3}\, t + 0.160 \times 10^{-6}\, t^2$ (t = 25 to 1300°C), calculated by the authors of this article from the data in [5];

$a \text{ (in pm)} = 508.4 + 4.260 \times 10^{-3}\, t + 1.280 \times 10^{-6}\, t^2$ (t = 0 to 1170°C), calculated by the authors of this article from the data in [7],

where the lattice expansion coefficient for the last equation is given by $1/a(25°C) \cdot da/dt = 8.38 \times 10^{-6} + 5.03 \times 10^{-9}\, t$ (t = 0 to 1000°C). Above 1000°C, the lattice parameter of N-saturated α-Th increases more rapidly than that of pure α-Th due to expansion by dissolved N. Above 1000°C it is represented to within 0.15% by the equation

$$a \text{ (in pm)} = 522.3 - 21.06 \times 10^{-3}\, t + 12.78 \times 10^{-6}\, t^2 \text{ (t = 1000 to 1605°C) [7]}.$$

References for 4.1.2.4 see p. 9

This equation gives a values for α-Th coexisting with ThN and shows a strong upward curvature in its temperature dependence. The curve ends at the α → β transition temperature of 1605°C with the value a = 521.4 pm as compared to the value a = 518.5 pm for the hypothetical superheated pure α-Th as obtained by extrapolation of the last two equations (see p. 7). Measurement of the temperature dependence of a is complicated by a strong tendency to grain growth. The lattice parameters of β-Th at the α → β transformation and at the α-Th–β-Th–ThN peritectoid temperatures were reported by Chiotti, White [5]. The a value for β-Th, 412.4 pm at 1605°C, was obtained by extrapolation of the values for pure β-Th at lower temperatures [5] assuming that any effect of N solubility was negligible. The following additional values are given by Chiotti, White [5] for β-Th: 411.5 (1425°C), 411.7 (1455°C), and 411.9 ± 0.2 pm (1500°C). The solubility of N in β-Th begins to increase rapidly above 1700°C, and a strong increase in the lattice parameter of β-Th coexisting with ThN over that of pure β-Th is to be expected.

Mechanical Properties

The density of N-saturated α-Th and β-Th at 1605°C is calculated by the authors of this article from the above lattice parameter data as 10.93 and 10.99 g/cm^3, respectively.

Additions of N in amounts of up to 0.4 wt% (= 6.3 at%) N induce a linear variation in the following mechanical properties of α-Th at room temperature [9, 10]:

        a) increase of the tensile strength from 190 to 230 MN/m^2,

        b) increase of the yield strength from 110 to 130 MN/m^2,

        c) increase of the Vickers hardness from 65 to 78, and

        d) decrease of the area reduction from 60 to 50%.

Plastic flow characteristics of polycrystalline α-Th containing 0.04 to 0.97 at% (600 ppm) N have been reported by McLachlan [6] and Peterson, McLachlan [11]. Tensile flow stresses at 1% strain, σ, were measured at 4.2 to 800 K under conditions of constant flow rates, 5×10^{-5} to 10^{-3} s^{-1}, and creep strain rates of 10^{-11} and 10^{-5} s^{-1} at constant stress. The N in solid solution increases the alloy strength and resistance to flow as measured by σ, which can be resolved into two components as follows:

1) σ*, a thermally activated component, which decreases with increasing temperature up to the critical temperature, 360 K, where it vanishes, varies as $\sqrt{x_N}$ at 4.2 and 77 K, and is proportional to x_N at 198 to 333 K; and

2) $σ_\mu$, an athermal component, independent of dissolved N and temperature with the mean value of 46 MN/m^2. Above the critical temperature, the apparent σ may increase to values greater than $σ_\mu$ due to aging (ThN precipitation).

Nitrogen in solid solution with Th is postulated to increase the strength and flow stress by acting as a strong short-range barrier to dislocation motion. Observed increase in flow stress due to strain hardening is shown to be athermal in character, independent of the N content. Strain-rate sensitivity of the alloys is expressed by the strain rate parameter, $\partial σ/\partial \ln \dot{\varepsilon}$, which increases with N content and varies with temperature, showing two peaks, one at 273 K associated with dissolved N. The second peak occurs in the two supersaturated alloys with 350 and 600 ppm N at 623 K and is due to age-hardened structures that develop during the tensile testing. Like σ*, $\partial σ/\partial \ln \dot{\varepsilon}$ varies as $\sqrt{x_N}$ at 77 K and as x_N at 196 to 473 K [6, 11].

Nitrogen additions were found to reduce the activation volume for flow but, for a given x_N, it is constant up to 273 K where the strain rate parameter is a maximum and then increases at higher temperatures. The latter increase is attributed to a decrease in the effective number of activated obstacles having a range of strengths. Nitrogen was found to have no significant effect on the flow activation energy, which increases with temperature to 1.3 eV at the critical temperature. Age-hardening was observed with the high-N, 350- and 600-ppm, alloys. The maximum increase in age-hardening by annealing of these quenched alloys was found to appear at 573 K where the diamond pyramid hardnesses increased from 57 to 93 in the 350- and from 78 to 130 in the 600-ppm alloy with constant strain hardening [11].

Transmission electron microscopy shows that the size of the ThN particles in aged alloys was 100 to 550 Å at 673 K, 50 to 450 Å at 623 K and < 50 Å when aged at the temperature of maximum hardness, 573 K [11].

Electrical Resistivity

Incorporation of N into α-Th at concentrations in the range of 0.04 to 1 at% N increases the electrical resistivity at 4.2 K. The results of measurements as read from a plot can be represented by the equation $R_{4.2K}$ (in $\mu\Omega \cdot m$) = $33 + 21.2 \times 10^3 x_N$ (N atom fraction x_N = 0.0004 to 0.01). The alloys after water-quenching from 1000°C obey this equation at compositions up to x_N = 0.01, but not if they are furnace-cooled from 800°C, because under these conditions some of the N can precipitate out of solution. The latter alloys with $x_N > 0.004$ show a resistance plateau at $R_{4.2K}$ = 115 $\mu\Omega \cdot m$ [6].

References for 4.1.2.4:

[1] Benz, R., Hoffmann, C. G., Rupert, G. N. (J. Am. Chem. Soc. **89** [1967] 191/7).
[2] Burgers, W. G., van Liempt, J. A. M. (Z. Anorg. Allgem. Chem. **193** [1930] 144/60).
[3] James, W. J., Straumanis, M. E. (Acta Cryst. **9** [1956] 376/9).
[4] Blumenthal, B., Sanecki, J. E. (J. Nucl. Mater **22** [1967] 100/2).
[5] Chiotti, P., White, R. W. (J. Nucl. Mater. **23** [1967] 37/44).
[6] McLachlan, D. R. (Diss. Iowa State Univ., Ames, Iowa, 1973).
[7] Benz, R., Balog, G. (High Temp. Sci. **3** [1971] 511/22).
[8] Harris, I. R., Raynor, G. V. (J. Less-Common Metals **7** [1964] 11/6).
[9] Milko, J. A., Adams, R. E., Harms, W. O. (Metal Thorium Proc. Conf., Cleveland 1956 [1958], pp. 186/216).
[10] Goldhoff, R. M., Ogden, H. R., Jaffee, R. I. (BMI-776 [1952] 1/44; N.S.A. **10** [1956] No. 2715).

[11] Peterson, D. T., McLachlan, D. R. (Met. Trans. A **6** [1975] 1359/66), Peterson, D. T., McLachlan, D. (IS-2600 [1971] 1/312, 142; N.S.A. **26** [1972] No. 17467).

4.1.2.5 Chemical Analyses

Thorium. The Th contents of binary mixtures are determined simply by ignition in a muffle furnace at 800 to 900°C. The product is weighed as ThO_2. Calculation of nitrogen by difference, however, can lead to anomalously high apparent N contents when oxygen, a common impurity, is present in appreciable amounts.

The several methods of chemical analysis for Th can be classified as follows [1, 2]:

Gravimetric. Th^{4+} is precipitated as a hydroxide, peroxide, oxalate, or other compound and ignited to ThO_2.

Volumetric. Th^{4+} is precipitated as a molybdate, iodate, peroxide, oxalate, or other compound and titrated [3 to 5].

Spectrophotometric. Measurements of the color intensity formed by reaction of Th^{4+} with organic reagents such as Thoron [6], Arsenazo, Morin [7, 8] and others [9 to 14].

With binary Th–N alloys having no interferring cation, the spectrophotometric methods lend themselves to rapid routine analyses especially useful for limited sample size, but the gravimetric techniques are ordinarily most accurate.

Nitrogen. The Kjeldahl method or one of its modifications is most commonly used. The sample is digested to convert the N to NH_4^+ and distilled from an alkaline solution as NH_3, absorbed in standard H_2SO_4 solution and the excess titrated with a standard base or absorbed in a boric acid solution and titrated with standard 0.1N H_2SO_4. Digestion techniques have been reviewed [15, 16] and recommendations made [17]. Other modifications of the Kjeldahl method are 1.) to fuse the sample in pre-fused KOH or NaOH at 500 to 600°C and scrub the off-gases in hot 20% NaOH solution in an Ar stream instead of digesting [18], 2.) instead of titrating to determine the NH_3 photometrically using the Nessler's reagent [19] or indophenol [15, 20], or 3.) to digest the sample in a 1:1 mixture of Cu selenate and hydrofluosilicic acid in HCl solution [17]. A colorimetric technique applicable to binary nitrides and which avoids the distillation step uses a pyridine-pyrazolone reagent [21]. Vacuum fusion techniques are not suited for determination of N in Th alloys [22, 23]. Modified Dumas techniques which have been used are to fuse the sample with $Na_2S_2O_7$ [24], $PbO + PbCrO_4$ [25], or Co_3O_4 [26].

Minor Components

Carbon. Analyses are conventionally done by heating the sample to 1200°C in a stream of O_2 in a combustion tube. The product gases are carried through CuO to convert CO to CO_2, through MnO_2 or $Cr_2O_3 + H_2SO_4$ solution for purification, P_2O_5 column for drying, and are finally absorbed on asbestos impregnated with NaOH (Ascarite) and weighed, or the CO_2 is determined volumetrically.

Oxygen. Analyses are ordinarily made by vacuum fusion with Fe or Pt [22, 23, 27] or by inert gas fusion [28]. The sample is dissolved in the molten metal saturated with carbon in a graphite crucible at 1900°C and the evolved CO converted to CO_2 in a CuO column and measured gas-volumetrically. More reproducible results are said to be obtained by neutron activation analysis for oxygen. With this method the sample is exposed a few minutes to 14 MeV neutrons to activate the oxygen by the nuclear reaction $^{18}O(n,\gamma)^{19}O$ and the amount of oxygen is obtained from the ^{19}O activity having the half life of $t_{1/2} = 29.1$ s [29] by comparison with that of a standard. The technique of isotopic dilution which has been used to analyze for oxygen in Ti should be applicable too [30].

References for 4.1.2.5:

[1]   Rodden, C. J., Warf, J. C. (Natl. Nucl. Energy Ser. Div. VIII 1 [1950] 160/207).
[2]   Ryabchikov, D. I., Gol'braikh, E. K. (Analytical Chemistry of Thorium, Humphrey, Ann. Arbor, Mich., 1969, pp. 1/289, 16/73).
[3]   Banks, C. V., Diehl, H. (Ind. Eng. Chem. Anal. Chem. **19** [1947] 222/4).
[4]   Flaschka, H., ter Haar, K., Bazen, J. (Mikrochim. Acta **1953** 345/8).
[5]   Körbl, J., Přibil, R., Emr, A. (Collection Czech. Chem. Commun. **22** [1957] 961/6).
[6]   Grimaldi, F. S., Fletcher, M. H. (Anal. Chem. **28** [1956] 812/6).
[7]   Blank, A. B., Mirenskaya, I. I., Satanovskii, L. M. (Zh. Analit. Khim. **30** [1975] 1116; J. Anal. Chem. [USSR] **30** [1975/76] 939/43).
[8]   Fletcher, M. H., Milkey, R. G. (Anal. Chem. **28** [1956] 1402/7).

[9] Charlot, G. (Colorimetric Determination of Elements, Elsevier, New York 1964, pp. 399/402).
[10] Fritz, J. S., Ford, J. J. (Anal. Chem. **25** [1953] 521/2).

[11] Lange, B., Vejdelek, Z. J. (Photometrische Analyse, Verlag Chemie, Weinheim 1980, pp. 325/30).
[12] Reilley, C. N., Barnard Jr., A. J., Püschel, R. (in: Meites, L., Handbook Analytical Chemistry, McGraw-Hill, New York 1963, pp. 3/77).
[13] Sandell, E. B. (Colorimetric Determination of Traces of Metals, 3rd Ed., Intersci., New York 1959, pp. 1/1054).
[14] ter Haar, K., Bazen, J. (Anal. Chim. Acta **9** [1953] 235/40).
[15] Boltz, D. F., Taras, M. J. (Nitrogen, in: Boltz, D. F., Howell, J. A., Colorimetric Determination of Nonmetals, Chapter 7, Wiley, New York 1978, pp. 197/251).
[16] Kirk, P. L. (Anal. Chem. **22** [1950] 354/8).
[17] Lathouse, J., Huber Jr., F. E., Chase, D. L. (Anal. Chem. **31** [1959] 1606/7).
[18] Passer, R. G., Hart, A., Julietti, R. J. (Analyst **87** [1962] 501/3).
[19] Jensen, K. J., Mundy, R. J. (Natl. Nucl. Energy Ser. Div. VIII **1** [1950] 208/17; C.A. **1951** 1912).
[20] Bohnstedt, U. (Z. Anal. Chem. **163** [1958] 415/22).

[21] Kruse, J. M., Mellon, M. G. (Anal. Chem. **25** [1953] 1188/92).
[22] Booth, E., Bryant, F. J., Parker, A. (Analyst [London] **82** [1957] 50/61).
[23] Griffith, C. B., Albrecht, W. M., Mallett, M. W. (BMT-1033 [1955]).
[24] Düsing, W., Hüniger, M. (Tech. Wiss. Abhandl.-Osram-Ges. **2** [1931] 357/65).
[25] Satoh, S. (Sci. Papers Inst. Phys. Chem. Res. [Tokyo] **35** [1939] 182/90).
[26] Kusakabe, T., Imoto, S. (Technol. Rept. Osaka Univ. **22** [1972] 477/88).
[27] Taylor, R. E. (Anal. Chim. Acta **21** [1959] 549/55).
[28] Smiley, W. G. (Anal. Chem. **27** [1955] 1098/102).
[29] Bujdosó, E., Fehér, I., Kardos, G. (Activation and Decay Tables of Radioisotopes, New York 1973, p. 25).
[30] Masson, C. R., Whiteway, S. G., Jamieson, W. D., Collings, C. A. (Can. Met. Quart. **5** [1966] 329/41).

4.1.3 Thorium Mononitride ThN

4.1.3.1 Preparation

General Remarks

The most direct preparations of the binary compounds depend, in principle, on reaction of N_2 with the metal. Because there is virtually no self-purification by preferential vaporization, the starting metal purity should be at least as good as that required in the endproduct. Thorium metal is produced commercially by Ca reduction of ThO_2 (or ThF_4) followed by compaction and arc melting [1]. Typical analyses of such metal lists the major impurity as 1000 to 2000 ppm oxygen [2] corresponding to 0.7 to 1.4 mol% ThO_2. Purification can be made electrolytically using fused $ThCl_4 + KCl + LiCl$ (or NaCl) as electrolyte [3] but probably the most consistently pure metal is produced by the van Arkel-deBoer iodide process [4]. This quality of metal has been available commercially as "Crystal-Bar" Th since the mid fifties. The upper impurity level

References for 4.1.3.1 see pp. 14/5

of "Crystal-Bar" Th was given in 1958 as 50 ppm oxygen, 20 ppm N, and 40 ppm C [2]. These three elements have an inordinate influence on some boundary compositions of Th–N phases. Purification by electrotransport has also been reported [5].

Direct Nitridation of Thorium

Reaction of massive Th metal with 1 atm N_2 at temperatures of 670 to 750°C leads to a very thin film of ThN [6] and at higher temperatures up to 1490°C to an inner ThN together with an outer gray, presumably Th_3N_4, film. Reaction of N_2 at pressures below the Th_3N_4 decomposition pressure (compare the "Th_3N_4–ThN–N_2 equilibria", p. 39) and at temperatures between 1400 and 1700°C gives porous ThN layers [8], but the reaction is too slow as a preparative method.

Preparation of ThN by reaction of 2 atm N_2 with Th melted in an induction-heated tungsten crucible has been reported by Olson, Mulford [9]. This technique is only suited for small preparations if the expensive crucible is not to be sacrificed. The crucible should be heated slowly so as to avoid direct contact of liquid Th metal with the W which alloys rapidly.

High-density ThN + Th alloys have been prepared by arc-melting of Th in 1 atm N_2. Free Th present in the product in amounts of ca. 3% is subsequently eliminated by annealing the alloy in 0.5 atm N_2 for 6 h at 2400°C [10 to 12]. The final nitride product with 3-mm grain size contains some nonuniformly distributed porosity. Thümmler et al. [13] have made high density ThN buttons by arc-melting of Th in 15 atm N_2 either with a permanent or consumable electrode. The product ThN contained no free Th but did have a minor amount of the higher nitrides.

Dense ThN has been made by multiple zone melting of a Th rod in 1.3 atm N_2 and eliminating the small amount of free Th by a post-anneal for 6 h in 0.5 atm N_2 at 2350°C [12].

Direct formation of ThN powder has been made by throttled nitridation of Th powder at 1600°C [14], but the preparation is usually made in two stages with Th_3N_4 as intermediate as described below.

Th_3N_4 Decomposition

The Th_3N_4 powder formed by reaction of excess N_2 with finely divided metal or ThH_4 can be decomposed to ThN in vacuum with continuous pumping at temperatures above about 1200°C. The product is compacted to a more or less porous ThN body by powder metallurgical techniques.

Thorium metal turnings react with excess N_2 at 400 to 940°C and the product Th_3N_4 decomposes to ThN powder in vacuum at 1500°C [15]. This technique can be used to convert commercial arc-melted round-bar Th derived from the Ca reduction process. It is a convenient preparative method when economy rather than purity of the product ThN powder is decisive.

Thorium powder is available commercially and has often been used to prepare ThN. The powder is reacted by slowly heating it in 1 atm N_2 to 800 to 900°C where the reaction is maintained for 1 to 3 h [16 to 19] or by reacting for 3 h in 1 atm N_2 at 1200°C [20 to 22, 27]. After cooling, the product Th_3N_4 powder is transferred to an Mo or, preferably, W crucible, evacuated, and slowly heated by induction to 1500°C (with an Mo crucible) or 1600°C (W crucible) under continuous pumping to the end pressure of 2×10^{-9} atm. This method has the advantage of shifting the onus of hydriding and dehydriding to a producer of powder, but, for best control of purity, the powder is prepared in the same apparatus used for the nitridation reaction with ThH_4 as an intermediate, as described next.

ThH₄ Nitridation

The reaction is made on a laboratory scale with a 30-mm diameter quartz tube long enough to extend through the hinged furnace but short enough to fit, subsequently, into the air-lock of a glove box. The tube is necked down at the gas exit end and closed with a stopcock. The 30-mm diameter inlet end is closed with a tapered joint and cap having the gas inlet through a stopcock. With the Th metal loaded at midlength, the evacuated quartz tube is heated in the split-tube furnace to 400°C and the reaction with H_2 allowed to proceed 8 to 20 h. The product is a voluminous finely divided bluish-black Th_4H_{15} or ThH_4 powder. Initiation of the reaction, especially with less pure metal, is sluggish. In such cases, the initiation can usually be secured by a brief heating of the metal to 800°C using a torch played on the outer wall of the quartz tube. The ThH_4 is converted to Th_3N_4 in a stream of N_2 by slowly heating to 900°C [23 to 26]. The product, brown Th_3N_4 powder, is transferred in a dry inert atmosphere to a W crucible, evacuated, and slowly heated under continuous pumping to 1600°C. When a vacuum of 1.3×10^{-9} atm is reached, the decomposition is complete. The decomposition is slower but has been reported to be effected at the following lower temperatures: 1500 [26], 1400 [17, 22], 1350 [25, 27], and 1200 to 1350°C [21, 24, 28]. The product yellow powder is readily cold-pressed to a self-supporting shape and can be sintered in vacuum to about 90% theoretical density. Prolonged sintering above 1800°C in a hard vacuum enhances the bulk density somewhat by formation of intergranular liquid Th which can be back-reacted by annealing in N_2.

Reaction of ThBr₄ with KNH₂

Reactions occurring in liquid NH_3 solutions in the presence of excess KNH_2 and K are as follows [29]:

$$ThBr_4 + 3KNH_2 + NH_3(l) = K_2ThN_2 \cdot KNH_2 + 4HBr \text{ and}$$
$$ThBr_4 + 5K + 3NH_3(l) = KThN(NH) \cdot NH_3 + 4KBr + \tfrac{5}{2}H_2.$$

From the product salts, ThN is obtained by thermal decomposition of the imide radical and separation of the K metal by vacuum distillation.

Reaction of K₂Th(NO₃)₆ with KNH₂

Liquid NH_3 solutions of anhydrous $K_2Th(NO_3)_6$ and 5.2 to 7 moles of KNH_2 are reacted by mixing. The white precipitate, $KTh(NH)_x(NH_2)_{5-2x}$, thus obtained is separated and heated in a stream of N_2 to 270°C to form $K_3Th_3N_5$ which, in turn, decomposes at 290 to 425°C to amorphous ThN [30].

Nitridation of Thorium Oxalate

A patent has been issued for the preparation of pure ThN or its mixture with either U, Pu, or any other of a number of the high-melting transition metal nitrides [31]. To prepare pure ThN, an intimate fine powder mixture of C with Th oxalate in the mole ratio of 1 to 2 is heated in streaming H_2 or $Ar + H_2$ to above 600°C (the oxalate decomposes at 750 to 800°C) after which the product is evacuated and the carbothermal reduction completed at 1100 to 1300°C. The product is converted to the nitride in streaming N_2 by heating to 1750 to 1800°C until evolution of CO is complete.

Preparation and Formation of Special Forms of ThN

Microspheres. A patent has been issued for the formation of microspheres of U nitrides by a process claimed to be applicable for the formation of Th nitride microspheres [32]. In the described process, U or $UO_2 + C$ microspheres are made by a sol-gel process, vacuum dried, heated in Ar to 1700°C and 16 h at 1500°C in streaming H_2 and finally converted to the nitride in streaming N_2. Another patent has been issued for the formation of microspheres with Th nitride coatings [7]. In this process core particles consisting of nitrides of U, Pu, or Th have the coating deposited during exposure to reacting gaseous ThF_4 and NH_3 in a fluidized bed reactor at 600 to 1000°C.

ThN Layers. Layers of ThN formed on solid Th(1400 to 1700°C) in N_2 at pressures below the Th_3N_4 decomposition pressure are porous [8]. Growth of such scales is, therefore, not diffusion-controlled. The initial ThN film is evidently partly protective, but, after a period of time breaks down from stresses induced by the reaction volume changes, and a new film then forms over the freshly exposed metal surface as suggested by the account given by Gerds, Mallett [6].

Reaction of N_2 with liquid Th spheres has been investigated. Five-mm diameter spheres of Th supported on a bed of ThN powder grow a dense protective scale of ThN when reacted with an N_2 atmosphere [8]. For the kinetics of the diffusion-controlled growth of scale thickness, see p. 24.

References for 4.1.3.1:

[1]   Cuthbert, F. L. (Thorium Production Technology, Chapter 6, "Preparation of Th metal by reduction", Addison-Wesley, Reading, Mass., 1958, pp. 146/201).
[2]   Wilhelm, H. A., McCarley, R. E. (ISC-1029 [1958] 1/52; N.S.A. **13** [1959] No. 10022).
[3]   Noland, R. A. (Metal Thorium Proc. Conf., Cleveland 1956 [1958], pp. 124/32).
[4]   Veigel, N. D., Sherwood, E. M., Campbell, I. E. (Metal Thorium Proc. Conf., Cleveland 1956 [1958], pp. 104/13).
[5]   Peterson, D. T., Schmidt, F. A., Verhoeven, T. D. (Trans. AIME **236** [1966] 1311/5).
[6]   Gerds, A. F., Mallett, M. W. (J. Electrochem. Soc. **101** [1954] 175/80).
[7]   Naumann, D., Burk, W., Heyne, W. (Fr. 1489718 [1967] from C.A. **68** [1968] No. 110699).
[8]   Benz, R. (J. Electrochem. Soc. **119** [1972] 1596/602).
[9]   Olson, W. M., Mulford, R. N. R. (J. Phys. Chem. **69** [1965] 1223/6).
[10] Chang, J. Y., Ewing, W. M., Fulkerson, W., McElroy, D. L., Weaver, S. C. (U.S. 3804928 [1974]).

[11] Fulkerson, W., Kollie, T. G., Weaver, S. C., Moore, J. P., Williams, R. K. (Plutonium 1970 Other Actinides Proc. Intern. Conf., 1970, pp. 374/85).
[12] Weaver, S. C. (Diss. Univ. Tenn., Knoxville 1972).
[13] Thümmler, F., Wedemeyer, H., Politis, C. (KFK-1023 [1969] 1/14; N.S.A. **24** [1970] No. 12454).
[14] Ettmayer, P., Waldhart, J., Vendl, A., Banik, G. (Monatsh. Chem. **111** [1980] 945/8).
[15] Chiotti, P. (J. Am. Ceram. Soc. **35** [1952] 123/30).
[16] Aronson, S., Auskern, A. B. (J. Phys. Chem. **70** [1966] 3937/41).
[17] Auskern, A. B., Aronson, S. (J. Phys. Chem. Solids **28** [1967] 1069/71).
[18] Aronson, S., Cisney, E., Gingerich, K. A. (J. Am. Ceram. Soc. **50** [1967] 248/52).
[19] Kamegashira, N., Tsuji, T., Miyamoto, T., Naito, K. (J. Nucl. Mater. **102** [1981] 26/9).
[20] Kusakabe, T., Imoto, S. (Nippon Kinzoku Gakkaishi **35** [1971] 795/800).

[21] Kusakabe, T., Imoto, S. (Technol. Rept. Osaka Univ. **22** [1972] 477/88).
[22] Sugihara, S., Imoto, S. (J. Nucl. Sci. Technol. [Tokyo] **8** [1971] 630/6).
[23] Benz, R., Hoffman, C. G., Rupert, G. N. (J. Am. Chem. Soc. **89** [1967] 191/7).
[24] deNovion, C. H. (CEA-R-4113 [1970] 1/153; N.S.A. **25** [1971] No. 11980).

[25] Ono, F., Kanno, M., Mukaibo, T. (J. Nucl. Sci. Technol. [Tokyo] **10** [1973] 391/5).

[26] Ozaki, S., Kanno, M., Mukaibo, T. (J. Nucl. Sci. Technol. [Tokyo] **8** [1971] 41/4).

[27] Adachi, H., Imoto, S. (Technol. Rept. Osaka Univ. **23** [1973] 1121/54, (Eng.) 425/9).

[28] Kusakabe, T., Imoto, S. (Nippon Kinzoku Gakkaishi **35** [1971] 1115/20).

[29] Young, R. C. (J. Am. Chem. Soc. **57** [1935] 1195/6).

[30] Schmitz-Dumont, O., Raabe, R. (Z. Anorg. Allgem. Chem. **277** [1954] 297/314).

[31] Wedemeyer, H. (Ger. 2262868 [1978]).

[32] Triggiani, L. V. (Fr. 1526427 [1968] from C.A. **71** [1969] No. 18036).

4.1.3.2   Crystallographic and Structural Properties

Crystallographic Data

Thorium mononitride is a stoichiometric phase at all temperatures; see Section 4.1.2.3, p. 6.

ThN crystallizes in a face-centered cubic NaCl-type structure with the space group Fm3m (No. 225) [1, 2].

The reported lattice parameter a falls mostly in the range of values 515.6 to 516.2 pm, corresponding to the densities of 11.92 to 11.88 g/cm^3, and are summarized in Table 2. Olson, Mulford [3] have measured the lattice spacing by the Debye-Scherrer method of ThN that had been subjected to a number of different thermal treatments. Their reported values a = 515.8 to 516.13 pm appear to be independent of whether the ThN was quenched from the melt or vacuum-annealed at 2000 or at 1000°C. Their most pure ThN was probably that prepared *in situ* and melted in 1.5 atm N$_2$ giving a = 515.85 ± 0.03 pm (row 14 and 15 of Table 2). Annealing this product at 2000°C had no significant effect on the a value. Other melted specimens containing up to 1.2% oxygen gave values of a = 515.90 ± 0.03 pm (row 16). In another investigation [4], the residue of specimens of ThN(+O) after being partly distilled from a W crucible in the temperature range 1540 to 1930°C gave an unusually wide range of a values all measured by the Debye-Scherrer method, viz. 515.5 to 516.2 pm (row 22 of Table 2) [4]. The values tend to be higher in those specimens heated to higher temperatures above 1800°C, where there is a small but definite oxygen solubility in ThN [6, 19]. The evidence is that oxygen dissolves in the ThN lattice as ThO and that it can be frozen in the lattice by quenching and thereby influence the lattice parameters. In 6 coordination, O^{2-} ions are smaller than N^{3-} ions so that substitution of O for N should reduce the a value. The effect, however, appears to be small. A reduction of the a value by 0.1 to 0.3 pm by dissolved oxygen has been reported by Kusakabe, Imoto [6] (see rows 24 to 26 of Table 2).

Congruently melted ThN as prepared from iodide Th after quenching has the composition ThN$_{0.995 \pm 0.005}$, is single phase with a = 515.7 ± 0.1 pm (row 21 of Table 2). Subsequent vacuum-annealing at 1430°C has no effect on the a value but does coagulate excess Th into tiny islands of free Th, indicating that the excess Th must precipitate rapidly from the ThN lattice but is not optically resolvable upon quenching. Kusakabe, Imoto [5, 7] have reported the rather high value a = 518.0 pm for ThN coexisting with either Th$_3$N$_4$ or Th$_2$N$_2$O (row 1 of Table 2) after quenching from 1200 to 1350°C. This result has not yet been confirmed.

   References for 4.1.3.2 see p. 19

Table 2
Room-Temperature Lattice Parameter a of ThN.

No.	a in pm	secondary phases or oxygen impurity	thermal treatment			Ref.
			temperature in °C	atmosphere	time in h	
1	518.0	Th_3N_4 or Th_2N_2O	1200 to 1350	7×10^{-4} atm N_2		[5, 7]
2	515.6 to 516.1	with or w.o. ThO_2	1200 to 1350	vacuum		[5 to 7]
3	516.5 ± 0.2					[8]
4	516.3 ± 0.2		1877	vacuum		[8]
5	516.4 ± 0.2		1957	vacuum		[8]
6	516.3	ThN (free Th) + 0.3% O	1500	vacuum		[9]
7	516.4 ± 0.2	ThO_2	480	1 atm N_2		[9]
8	516.3		1400	vacuum	3	[10]
9	516.19 ± 0.02		1500	vacuum		[11]
10	515.94 ± 0.01		1600	vacuum	2	[12]
11	515.97 ± 0.07	free Th + 0.04% O	melted	2 atm N_2		[3]
12	516.08 ± 0.05	free Th + 0.04% O	1800	vacuum	1	[3]
13	515.9 ± 0.1	free Th + 0.04% O	1000	0.92 atm N_2	96	[3]
14	515.85 ± 0.03	free Th + ca. 0.04% O	melted	1.5 atm N_2		[3]
15	515.83 ± 0.03	free Th + ca. 0.04% O	2000	vacuum		[3]
16	515.90 ± 0.03	free Th + 0.6 to 1.2% O	melted	2 atm N_2		[3]
17	515.85 ± 0.05	sintered in graphite	melted	2 atm N_2		[3]
18	515.82 ± 0.03	sintered in graphite +0.6% O	1700	vacuum		[3]
19	515.9 ± 0.2		1400	vacuum		[13 to 15]
20	516.1 ± 0.2		1400	vacuum	4	[16]
21	515.7 ± 0.1		melted	0.03 to 3 atm N_2	1/4	[4]
22	515.5 to 516.2	0.1 to 0.4% O	1540 to 1930	vacuum	10 to 96	[4]
23	515.91 ± 0.03	1.5% ThO_2	1600	vacuum		[17, 18]
24	515.8 to 515.9		1400	vacuum		[6]
25	515.7	$ThN_{0.79}O_{0.19}$	1600	vacuum	17	[6]
26	515.5	$ThN_{0.71}O_{0.30}$	1800	vacuum	5	[6]
27	515.9	$ThN_{0.96}$	1200 to 1350	vacuum		[7]

In summary, the cause for the reported differences in the a values of ThN has not yet been unambiguously identified. Some of the variations seem to be real, but they are too large to be attributed to differences in stoichiometry. There is difficulty in isolating and separately characterizing the other possible sources of variation such as oxygen content, W contamination, and thermal history. The best average room-temperature a value of ThN is suggested

References for 4.1.3.2 see p. 19

[21, 27] to be 515.8 ± 0.1 pm. The corresponding density is D $(25°C) = 11.909 \pm 0.007$ g/cm³. The resulting interatomic distances are: Th–Th = 364.7 ± 0.1 (12×) and Th–N = 257.9 ± 0.1 (6×).

Gerward et al. [20] measured the pressure dependence of the ThN lattice spacing by the isostatic pressure method using synchrotron radiation and a diamond anvil. They describe their results with the equation

$$p = (B/B_o')[(a_o/a)^{3B_o'} - 1] \text{ (0 to 47 GPa)}$$

where $B_o = V_o(p/V)_T = 175 \pm 15$ GPa is the measured bulk modulus, $B_o' = 4.0 \pm 4$ is its mean pressure derivative, and a_o is the lattice parameter at $p = 1$ GPa. Alternatively, the data can be represented to within 1.8 pm by the equation

$$(a/a_o) = 1 - 2.00 \times 10^{-3} p + 18.4 \times 10^{-6} p^2 \text{ (0 to 47 GPa)}.$$

No structural phase transition was observed in the investigated pressure range [20].

The lattice parameters of ThN at elevated temperatures have been measured by the X-ray diffraction powder method in 4 different investigations. The results are summarized in the following equations, calculated by the authors of this article from the data in the cited literature:

$$a \text{ (in pm)} = 515.84(\pm 0.01) + 3.81(\pm 0.021) \times 10^{-3} t \text{ (t = 18 to 900°C)} \dots \dots \dots \quad (1)$$

with the linear thermal expansion coefficient $\alpha = \dfrac{1}{a(25°C)} \cdot \dfrac{da}{dt} = (7.39 \pm 0.042) \times 10^{-6}$ K⁻¹ (0 to 900°C) [12];

$$a \text{ (in pm)} = 515.4 + 4.26 \times 10^{-3} t \text{ (t = 800 to 1300°C)} \dots \dots \dots \dots \dots \dots \dots \dots \dots \quad (2)$$

with the expansion coefficient $\alpha = 8.2 \times 10^{-6}$ K⁻¹ (t = 800 to 1300°C) [14];

$$a \text{ (in pm)} = 515.7 + 3.43 \times 10^{-3} t + 0.760 \times 10^{-6} t^2 \text{ (t = 25 to 2340°C)} \dots \dots \dots \dots \quad (3)$$

with the expansion coefficient α (in K⁻¹) = $6.65 \times 10^{-6} + 2.95 \times 10^{-9} t$ (t = 25 to 2340°C) [22];

$$a \text{ (in pm)} = 516.0 + 3.86 \times 10^{-3} t + 0.287 \times 10^{-6} t^2 + 51.9 \times 10^{-15} t^4 \text{ (t = 0 to 2000°C)} \dots \dots \quad (4)$$

with $\alpha = 7.48 \times 10^{-6} + 1.113 \times 10^{-9} t + 0.402 \times 10^{-15} t^3$ (t = 0 to 2000°C) [23].

In the temperature range of overlap, 800 to 900°C, equation (2) is lower by 0.2 pm while equation (3) is higher by 0.1 to 0.2 pm than equation (1). Equations (3) and (4) agree to within 0.2%.

The ThN compound is remarkable for having unusually large interatomic distances, and the Th atoms do not have the full 4 valence characteristic of many other Th–N compounds. Allbutt, Dell [25] have analyzed interatomic distances in the fcc monocompounds of the type MX formed between M denoting Th, U, Np, or Pu and the various nonmetal atoms, X, representing the main group V and VI elements of the periodic table. Taking into account the available magnetic susceptibility data, they show that, except for ThN, all the bond distances can be interpreted in terms of a single set of ionic radii in which the U ions exhibit the different radii, 87 or 93 pm, depending on the state of ionization, 5+ or 4+, respectively. The Th ion is assumed to have the fixed valence of 4+ and the ionic radius of $r_{Th^{4+}} = 99$ pm [24]. To bring the ThN anomaly into perspective, the bond distance data as reviewed by Allbutt, Dell [25] are compared with data on the monocarbides and hypothetical monoxides. For each different nonmetal X, the effective ion size and bond distance is different. The differences are given by the quantity $\Delta D = (D(Th-6X) - D(M-6X))$, where M-6X emphasizes that the M atom has a coordination number of 6 to the anion X. The ΔD values are all positive showing that the Th ion radius is always larger than the other M ion radii. The ΔD values decrease monotonically almost linearly from 19 for X = C to 13 for X = N to 6 pm for X = O, and show a further slight decrease with increasing size of X as X ranges down the main group V and VI elements of the periodic table. Allbutt, Dell [25] showed that the ΔD variations can be accounted for in terms of accepted cation radii for all the fcc monocompounds formed between M and the elements X (X = main group V and VI elements) except for N in ThN. The slightly larger ΔD values of the nitrides and the much larger values of the carbides must be attributed to an increased metallic character of the chemical bonds. This change in electronic structure must also be responsible for the reduced force constants in ThN and ThC [26] and the increased solubility in metallic Th. The O solubility is virtually nil, that of N is appreciable, and C forms a continuous series of solid solutions up to ThC.

Phonon Spectra

The phonon spectrum of ThN powder at 290 K has been determined from the inelastic scattering of 2-Å neutrons with a time-of-flight spectrometer at angles of 17° and 63°. The spectrum as illustrated in **Fig. 3** shows an optical peak at 10.3×10^{12} Hz. The corresponding force constant for a linear N oscillator is calculated as $A = \frac{1}{2} mw^2 = 48.7 \times 10^3$ dyn/cm = 48.7 N/m = 14.5 $e^2/2V$, where m is the mass of the N atom, w is the optical (phonon scattering) frequency, e is the electron charge, V is the volume per ThN unit in the crystal. The force associated with the constant A is less stiff than that of the analogous UN crystals with $A = 63.9 \times 10^3$ dyn/cm = 63.9 N/m = 16.2 $e^2/2V$. This can be attributed to the greater metallic character and to the greater interatomic distances in the ThN lattice. An additional weaker optical peak at 22×10^{12} Hz and an acoustical one at ca. 3.4×10^{12} Hz can be seen in Fig. 3 [26].

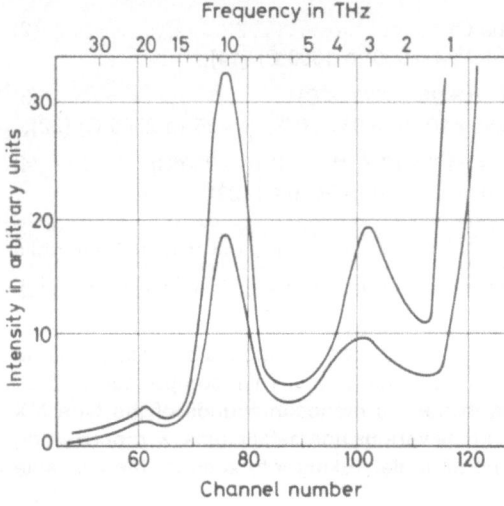

Fig. 3

Time-of-flight spectrum of ThN with 2-Å neutrons at the scattering angles 17° (lower curve) and 63° (upper curve). Three peaks are seen at 22×10^{12} (optical), 10.3×10^{12} (optical), and 3.4×10^{12} Hz (acoustic) [26].

Disorder

Balankin et al. [8] analyzed the stability of fcc $Th_{1-x}N_x$ crystals with ordered Th and N sublattices containing structural (nonstoichiometry) and thermal vacancies. Their results give for the free energy of formation of the structural vacancies

$$F_{Th^+} = 136\,200 + 25.0\ T\ (J/mol) = 32\,550 + 5.98\ T\ (cal/mol)$$
$$F_{N^+} = 488\,000 - 135.0\ T\ (J/mol) = 116\,600 - 32.3\ T\ (cal/mol)$$

and for the formation of thermal vacancies with the fractional vacant lattice sites of α at stoichiometry

$$RT\ln\ \alpha = -94\,800\ (J/mol) = -22\,660\ (cal/mol)\ [8].$$

References for 4.1.3.2:

[1] Chiotti, P. (AECD-3072 [1951] 1/63; N.S.A. **5** [1951] No. 3141).
[2] Rundle, R. E. (Acta Cryst. **1** [1948] 180/7).
[3] Olson, W. M., Mulford, R. N. R. (J. Phys. Chem. **69** [1965] 1223/6).
[4] Benz, R., Hoffman, C. G., Rupert, G. N. (J. Am. Chem. Soc. **89** [1967] 191/7).
[5] Kusakabe, T., Imoto, S. (Nippon Kinzoku Gakkaishi **35** [1971] 1115/20).
[6] Kusakabe, T., Imoto, S. (Nippon Kinzoku Gakkaishi **36** [1972] 305/9).
[7] Kusakabe, T., Imoto, S. (Technol. Rept. Osaka Univ. **22** [1972] 477/88).
[8] Balankin, S. A., Danilkin, E. A., Loshmanov, L. P., Skorov, D. M., Sokolov, V. S. (Met. Metalloved. Chist. Metal. No. 13 [1979] 59/63).
[9] Ozaki, S., Kanno, M., Mukaibo, T. (J. Nucl. Sci. Technol. [Tokyo] **8** [1971] 41/4).
[10] Sugihara, S., Imoto, S. (J. Nucl. Sci. Technol. [Tokyo] **8** [1971] 630/6).

[11] Venard, J. T., Spruiell, J. E., Cavin, O. B. (J. Nucl. Mater. **24** [1967] 245/6).
[12] Street, R. S., Waters, T. N. (AERE-M-1115; N.S.A. **17** [1963] No. 14804).
[13] Aronson, S., Auskern, A. B. (J. Chem. Phys. **48** [1968] 1760/5).
[14] Aronson, S., Cisney, E., Gingerich, K. A. (J. Am. Ceram. Soc. **50** [1967] 248/52).
[15] Aronson, S., Ingraham, A. (J. Nucl. Mater. **24** [1967] 74/9).
[16] Auskern, A. B., Aronson, S. (J. Phys. Chem. Solids **28** [1967] 1069/71).
[17] Danan, J. (CEA-R-4453 [1973]).
[18] Danan, J., deNovion, C. H., Dallaporta, H. (Solid State Commun. **10** [1972] 775/8).
[19] Benz, R. (J. Nucl. Mater. **43** [1972] 1/7).
[20] Gerward, L., Olsen, J. S., Benedict, U., Itié, J.-P., Spirlet, J. C. (J. Appl. Cryst. **18** [1985] 339/41), Olsen, J. S., Steenstrup, S., Gerward, L., Benedict, U., Itié, J.-P. (Rept. Univ. Copenhagen Phys. Lab. II, No. 85-14 [1985] 1/14; C.A. **103** [1985] No. 187192).

[21] Ackermann, R. J., Rauh, E. G. (J. Chem. Thermodyn. **4** [1972] 521/32).
[22] Benz, R., Balog, G. (High Temp. Sci. **3** [1971] 511/22).
[23] Politis, C. (KFK-2168 [1975] 1/86; C.A. **83** [1975] No. 200480).
[24] Zachariasen, W. H. (in: Seaborg, G. T., Katz, J. J., The Actinide Elements, Crystal Chemistry of the 5f Elements, McGraw-Hill, New York 1954, pp. 769/96).
[25] Allbutt, M., Dell, R. M. (J. Inorg. Nucl. Chem. **30** [1968] 705/10).
[26] Wedgwood, F. A. (J. Phys. **7** [1974] 3203/18), Wedgwood, F. A., deNovion, Ch. (AERE-R-7961 [1974] 66/73).
[27] Ackermann, R. J., Rauh, E. G. (High Temp. Sci. **4** [1972] 272/82).

4.1.3.3 Mechanical, Thermal, and Transport Properties

Density and Thermal Expansion

Some X-ray density data are given in the Section 4.1.3.2 "Crystallographic and Structural Properties" (see pp. 15/7). The thermal expansion is determined only by X-ray diffraction from the change in the lattice parameter; for details see p. 17.

Elastic Constants

The bulk modulus B_0 of ThN is 175 ± 15 GPa at room temperature [1]. With a calculated f occupational number of 0.9 (cf. "Band Structure", p. 31) the ThN bulk modulus $B_0 = 174$ GPa is deduced [2].

 References for 4.1.3.3 see pp. 25/6 2*

Thermal Decomposition, Vapor Pressure, Vaporization

The reaction of thermal decomposition in vacuum is $ThN(s) \rightarrow Th(s \text{ or } l) + \frac{1}{2}N_2$ [3, 4].

Balankin et al. [3] found that only the liquid Th appears under Langmuir vaporization conditions. The vapor partial pressures given as a function of stoichiometry are

$$\log p_{N_2} \text{ (p in Pa)} = 19.07 - 50\,800/T + 2\log (x/(1-2x))$$
$$\text{or}\quad \log p_{N_2} \text{ (p in atm)} = 14.06 - 50\,800/T + 2\log (x/(1-2x))$$
$$\text{and}\quad \log p_{Th} \text{ (p in Pa)} = 9.73 - 22\,820/T + \log ((1-2x)/(1-x))$$
$$\text{or}\quad \log p_{Th} \text{ (p in atm)} = 4.72 - 22\,820/T + \log ((1-2x)/(1-x)).$$

The equations describe the partial pressures as a function of x in $Th_{1-x}N_x$ and, in particular, define the congruent composition when the vapors have the same composition as the solid, i.e., when $p_{N_2}/p_{Th} = x/2(1-x)$.

Early measurements of N_2 partial pressures over ThN in a Ta Knudsen cell that were reported as being preliminary can be associated with the reaction $ThN(s) \rightarrow Th(s \text{ or } l) + \frac{1}{2}N_2$ and are given by the equation

$$\log p_{N_2} \text{ (p in atm)} = -18\,100/T + 2.89 \pm 0.05 \quad (T = 1810 \text{ to } 2280 \text{ K}) \text{ [5]}.$$

The solid material was analyzed by ignition as $ThN_{1.05}$, the apparent excess N being very probably due to oxygen present as impurity.

For pressures of N_2 coexisting with $ThN + Th(+N)(l)$ in the temperature range 2689 to 3093 K, see p. 22.

Balankin et al. [3] measured the rates of vaporization under the Langmuir conditions as being 0.59×10^{-4} and 1.78×10^{-4} kg·m^{-2}·s^{-1} at 2150 and 2230 K, respectively, which compares with the respective theoretical values of 0.38×10^{-4} and 1.34×10^{-4} kg·m^{-2}·s^{-1}. Their results show that ThN is stable in vacuum to 2190 ± 40 K as compared to the upper limit of 2073 K under Knudsen effusion conditions [4]. Taken in conjunction with the reported melting composition of $ThN_{0.995 \pm 0.005}$ [4], the results of Balankin et al. [3] establish that the solubility of Th in ThN is retrograde.

Vaporization of ThN and Th + ThN alloys from W Knudsen effusion-type crucibles in vacuum is extremely slow at temperatures below 1900°C [4]. The low rate of effusion has led to an erroneous interpretation of congruent vaporization. Re-analysis of the data is instructive. Some representative published data are listed in Table 3 [4]. The reported N/Th ratios as determined by ignition, disregarding the oxygen content, are listed in column 4. These results appear to show that the N/Th ratio of Th-rich specimens increases while that of N-rich (ThN) specimens decreases as if to a common composition previously projected to indicate a congruent vaporization composition. Correction of the compositions for the oxygen assumed to be present as ThO_2, column 6 of Table 3, shows, however, that two simultaneous reactions took place and governed the composition changes, viz. 1.) oxygen pickup under the vacuum conditions and 2.) preferential vaporization of N. The actual (N + O)/Th(nonmetal-to-Th) ratio will have some value intermediate between the values given in columns 4 and 6 of Table 3, but the conclusion is unchanged that the N/Th ratios all decrease monotonically with time (compare column 6 with 2 of Table 3). This shows unequivocally that the steady state vaporization is due to the decomposition reaction $ThN(s) \rightarrow Th(s \text{ or } l) + \frac{1}{2}N_2$.

The vaporization from the Th-rich specimens (rows 2, 5, and 7) is much slower than from more N-rich specimens. Vapors of Th are monatomic [6, 7]. According to the kinetic theory, the N_2 and Th molecules effuse at the relative rates

$$\frac{Z_{N_2}}{Z_{Th}} = \frac{p_{N_2}}{p_{Th}} \cdot \sqrt{\frac{232}{2 \times 14}} = 2.9 \frac{p_{N_2}}{p_{Th}} \approx 1.5$$

References for 4.1.3.3 see pp. 25/6

near congruent conditions. At 1800°C, $p_{Th} \approx 5 \times 10^{-9}$ atm, the rates of effusion of Th and N_2 vapors from Th + ThN mixtures through a 1-mm orifice are then about 2 mg Th and 0.4 mg N_2 per day. Thus, even if a congruent ThN phase were to exist, it is now evident that determination of the congruent composition by preferential Th vaporization from Th-rich alloys below 2000°C would be too slow for practical use. The data below 1600°C, the last 2 rows of Table 3, are inconclusive because of uncertainty about effects of dissolved oxygen and the vacuum background pressures. It is concluded that the partial pressure of any gaseous Th nitride is extremely low, $<10^{-9}$ atm near 2700 K, according to Gingerich [7]. Also, the existence of a gaseous Th nitride that would be required for congruency to exist is far too low for a congruent vaporization composition to fall in the ThN phase region.

Table 3

Results of Analyses on ThN and Th + ThN Specimens after Long-Term Vaporization from Knudsen-type W Crucibles [4].
Correction of the product N/Th ratios determined by ignition (column 4) after correction for oxygen assumed to be present as ThO_2 shows that all the N/Th ratios decrease monotonically with time (column 6) in accord with the reaction ThN(s) → Th(s or l) + ½N_2.

temperature in °C	initial N/Th ratio	time in h	product analyses			
			N/Th by ignition	% oxygen (% as ThO_2)	N/Th corrected for oxygen	phase composition
1901	1.0	19	0.91	0.35(2.9)	0.85	Th + ThN
1895	0.84	23	0.90	0.38(3.1)	0.83	Th + ThN
1805	1.0	70	0.92	0.31(2.6)	0.86	Th + ThN
1795	0.95	38	0.92	0.29(2.4)	0.87	Th + ThN
1657	0.88	78	0.91	0.26(2.1)	0.86	Th + ThN
1558	1.0	60	0.99	—	—	ThN
1537	0.97	94	0.99	0.27(2.2)	0.94	ThN

Pressures of N_2 in equilibrium with Th_3N_4 and ThN, representing the decomposition reaction $2Th_3N_4 \rightarrow 6ThN + N_2$, are treated in the Th_3N_4 Section (see p. 39).

Enthalpy of Formation, Gibbs Free Energy of Formation

The standard molar enthalpy of formation of ThN(s), $\Delta H_{f,298} = -88.5$ kcal/mol $= -370.3$ kJ/mol, is calculated from the available calorimetric data and N_2 decomposition pressures of Th_3N_4 as discussed on p. 39. This value is algebraically more positive than the previously given values -90.5 ± 2.5 [8, 9], -90.6 [10], -93.1 [11], and -93.5 kcal/mol [12]. Combined with the heat capacity data of Th_3N_4, cf. p. 40, the corresponding Gibbs free energy of formation of ThN at elevated temperatures is given by the equations (calculated by the authors of this article):

$$\Delta G_f \text{ (in cal/mol)} = -88\,000 + 20.24\,T \quad (T = 298 \text{ to } 2027 \text{ K}) \quad \dots \dots \dots \dots \quad (5)$$

$$\Delta G_f \text{ (in cal/mol)} = -91\,300 + 21.87\,T \quad (T > 2027 \text{ K}) \quad \dots \dots \dots \dots \dots \dots \quad (6)$$

The standard Gibbs free energy of formation of ThN can be formulated in terms of the partial free energies of the elements, each referred to their standard states, in the equilibrium reaction Th(s, l) + ½$N_2 \rightleftharpoons$ ThN as follows:

$$\Delta G_f(\text{ThN}) = G_{Th}^M + 0.5\,RT \ln p_{N_2} \quad \dots \dots \dots \dots \dots \dots \dots \dots \dots \dots \dots \dots \quad (7)$$

References for 4.1.3.3 see pp. 25/6

where, in the notation of Wagner [13], G_{Th}^M denotes the relative partial molar free energy of Th. Pressures of N_2 coexisting with $ThN + Th(+N)(l)$, have been reported by Olson, Mulford [14]. Their results are summarized in the equation

$$\log p \text{ (p in atm)} = 8.086 - 33224/T + 9.58 \times 10^{-18}\, T^5 \quad (T = 2689 \text{ to } 3093\text{ K)}.$$

At the temperatures of the measurements, there is quite appreciable solubility of N in Th(l) and estimation of the G_{Th}^M term is attended by an appreciable uncertainty. Estimation of ΔG_f by equation (7) is, however, useful for indicating the physical chemistry involved. Tentative rough estimates of $\Delta G_f(ThN)$ at some representative but arbitrary temperatures as summarized in Table 4 are based upon an extrapolation of equation (6) and the assumption of a regular Th–ThN solution. The estimates indicate that there must be a very strong interaction between the Th and N atoms in the liquid Th–N solution giving rise to strong negative deviations from ideal solution and that the p_{N_2} term must predominate in equation (7) at temperatures below 2689 K.

Table 4

Gibbs Free Energy of Formation of ThN as Approximated from Equation (7) with $G_{Th}^M =$ $RT \ln x_{Th(l)} - 50000 x_{N(l)}^2$ and Comparison with Values Obtained by Extrapolation of Equation (6). (The comparison indicates roughly the relative importance of the different terms in equation (7) and, in particular, shows that the p_{N_2} term must predominate at temperatures below 2689 K. A very strong Th–N interaction with negative deviations from ideality in the liquid Th–N solution appears probable.)

tempera-ture in K	$x_{Th(l)}$	$RT \ln x_{Th(l)}$ in cal/mol	p_{N_2} in atm	$\frac{1}{2}RT \ln p_{N_2}$ in cal/mol	$\Delta G_f(ThN)$ in cal/mol		
					equation (7)	equation (6) (extrapolation)	differ-ence
3093	0.50	−4300	2.6	+ 2900	−13800	(−23700)	9900
2689	0.78	−1300	1.2×10^{-3}	−18000	−21700	(−32500)	10800
(2300)	0.91	− 430	(1.8×10^{-6})	−30200	−31100	(−41000)	9900

The equation (7) is evaluated with the p_{N_2} data of Olson, Mulford [14], the liquidus compositions illustrated in Fig. 2, p. 5, and the assumption of regular liquid Th–N solutions with $G_{Th(l)}^M = RT \ln x_{Th(l)} - 50000\, x_{N(l)}^2$.

Gaseous Species. Gingerich [7] has identified nearly 100 ppm ThN(g) mass spectrometrically in vapors over ThP + BN mixtures contained in a W crucible at 2702 and 2745 K. From analysis of his data, Gingerich [7] gives the following enthalpies of reaction:

$$ThN(g) \rightleftharpoons Th(g) + \tfrac{1}{2}N_2: \ \Delta H_{298} = 23.1 \pm 6.0 \text{ kcal/mol},$$
$$ThN(g) \rightleftharpoons Th(g) + N(g): \ D_{298} = 138.0 \pm 8.0 \text{ kcal/mol},$$
$$ThN(s) \rightleftharpoons ThN(g): \ \Delta H_{298} = 203.1 \pm 10 \text{ kcal/mol, and}$$
$$ThN(s) + \tfrac{1}{2}N_2 \rightleftharpoons ThN(g): \ \Delta H_{f,\,298} = 112.5 \pm 7.5 \text{ kcal/mol}.$$

The 4 reactions are, respectively, 2 different modes of ThN(g) dissociation, ThN sublimation, and the formation of ThN(g) under standard conditions. In a review of these data, Rand [8] suggests the revised values $\Delta H_{298} = 26.8 \pm 6$ kcal/mol for the first reaction and $\Delta H_{f,\,298} = 116 \pm 6$ kcal/mol and gives for the Gibbs free energy of formation of ThN(g)

$$\Delta G_f(ThN(g)) = 111900 - 16.4\, T \text{ cal/mol}$$

with which he shows that ThN(g) is always a minor gaseous species over ThN.

References for 4.1.3.3 see pp. 25/6

Heat Capacity, Enthalpy, and Entropy

Reported molar heat capacity data are summarized in Table 5. deNovion [15] and deNovion, Costa [16] fitted their low temperature data over the temperature range 1.3 to 10 K to the empirical equation
$$C_v \approx C_p = \gamma T + \alpha T^3$$
to give for the temperature coefficient of the electronic heat capacity $\gamma = 3.12 \ mJ \cdot mol^{-1} \cdot K^{-2}$ and the coefficient of the lattice term as $\alpha = 84.7 \ \mu J \cdot mol^{-1} \cdot K^{-4}$. It follows from the equation $\gamma = \frac{1}{3}\pi^2 k^2 \ N(E_F) = 2.357 \ (mJ \cdot eV \cdot mol^{-1} \cdot K^{-2}) \ N(E_F)$ that the density of states at the Fermi level (including both spin directions) is $N(E_F) = 1.32 \ eV^{-1} \cdot atom^{-1}$. The coefficient of the electronic heat capacity of metallic Th is of the same order with the reported [20, 21] values $\gamma = 6.86$ [17], 4.34 [18], 4.31 [19], and 4.69 $mJ \cdot mol^{-1} \cdot K^{-1}$ giving in the same approximation $N(E_F) = 2.91$, 1.84, 1.83, and 1.99 $eV^{-1} \cdot atom^{-1}$ at the Fermi level, respectively. The Debye temperature associated with acoustic phonons, Θ_D, in ThN is calculated from the equation
$$\alpha = 12\pi^4 R/5\Theta_D^3 = 1943.6/\Theta_D^3 \ J \cdot mol^{-1} \cdot K^{-4}$$
as $\Theta_D = 284 \pm 2$ K. The above calorimetric data of deNovion, Costa [16] have been confirmed by Danan [22]. The lattice specific heat capacity of ThN was later resolved into two terms associated, respectively, with the acoustic (Debye function) and optic (Einstein function) phonons. The Einstein temperature $\Theta_E = 495 \pm 5$ K associated with optical phonons is reported by Danan et al. [23] and Danan [22].

From the temperature variation of the X-ray powder diffraction peak profiles of ThN, Aronson, Ingraham [28] calculate the characteristic Debye temperatures $\Theta_D = 234 \pm 15$ K on the basis of a "heavy-mass" model and $\Theta_D = 317$ K for an "average-mass" model. They give more credence to the former value in view of the large difference in masses between Th and N atoms and they claim that it agrees better with published entropy data.

High-temperature heat capacity data for ThN are summarized in rows 6 to 9 of Table 5. The data estimated by Rand [8] and given by the equation
$$C_p \ (in \ cal \cdot mol^{-1} \cdot K^{-1}) = 11.34 + 2.276 \times 10^{-3} \ T - 0.1144 \times 10^6 \ T^{-2}$$
are adopted here. Compared to this equation, the values estimated by Tagawa [26], row 7, are 8 to 17% higher. The equation given by Voitovich, Shakhanova [25], row 6, is based upon assumed Debye temperatures and melting point data and gives results 17% lower in value. The experimental data of Ono et al. [27], row 8, are higher by up to 7% at 1173 K.

Table 5

Summary of Heat Capacity Data for ThN.

| | method | temperature range in K | C_p (in $cal \cdot mol^{-1} \cdot K^{-1}$) = $a + b \times 10^{-3} T + c \times 10^6 T^{-2}$ | | | $S_{298.15}$ in $cal \cdot mol^{-1} \cdot K^{-1}$ | Ref. |
			a	b	c		
1)	adiabatic	1.3 to 10					[15]
2)	adiabatic	3 to 9					[16]
3)	adiabatic	7 to 300				13.4 ± 0.2	[23]
4)	adiabatic	5 to 300				13.4 ± 0.2	[22]
5)	adiabatic	4 to 350				13.7 ± 0.1	[24]
6)	estimated	300 to 2500	11.68	1.19	−0.131		[25]
7)	estimated	300 to 3000	tabulated				[26]

References for 4.1.3.3 see pp. 25/6

Table 5 (continued)

	method	temperature range in K	C_p (in cal·mol^{-1}·K^{-1}) = a + b×10^{-3} T + c×10^6 T^{-2}			$S_{298.15}$ in cal·mol^{-1}·K^{-1}	Ref.
			a	b	c		
8)	adiabatic	450 to 850	12.50	2.660	−0.2239		[27]
9)	estimated	298 to 2000	11.34	2.276	−0.1144		[8]

Based on their measurements of the low temperature heat capacities, the standard absolute entropy of ThN is reported as $S_{298}^\circ = 13.38 \pm 0.24$ cal·mol^{-1}·K^{-1} by Danan et al. [23] and Danan [22] and as 13.7 ± 0.2 cal·mol^{-1}·K^{-1} by Dell, Martin [24]. Following the precedent set by Rand [8], the latter value is adopted for consistency with data on Th$_3$N$_4$.

The following data are useful for converting reference states from 298.15 to 0 K [22, 23]:

$$H_{298} - H_0 = 2.020 \pm 0.036 \text{ kcal/mol,}$$
$$-(G_{298} - H_0) = 1.972 \pm 0.036 \text{ kcal/mol.}$$

Self-Diffusion

No measurements of self-diffusion coefficients of Th in the nitrides have been reported, but diffusion rates must be comparable to that in the analogous UN. Self-diffusion coefficients of ^{233}U in UN have been reported with the extremely low values as read from a plot ranging from 1×10^{-13} at 1873 K to 8.5×10^{-13} cm^2/s at 2133 K in 200 Torr N$_2$ and decreasing approximately linearly with N$_2$ pressure by a factor of 10^2 at 2 Torr [29].

The chemical diffusion of N in ThN has been studied in connection with the reaction of N$_2$ with liquid Th spheres. Five-mm diameter spheres of Th supported on a bed of ThN powder grow a dense protective scale of ThN when reacted with an N$_2$ atmosphere [30].

The growth of scale thicknesses, x_s, under these conditions increases as the square root of time. Values of the composition-average diffusion coefficients are calculated from the equation $\overline{D}_N(ThN) = x_s^2/4\ tu^2$ where the parameter u is the solution to the equation $\Delta X(ThN)/\Delta X(l,ThN) = \pi u\ e^{u^2}$ erf u, where $\Delta X(ThN)$ is the drop in N concentration across the ThN layer and $\Delta X(l,ThN)$ is the drop in N concentration at the inside interface between Th-saturated ThN and N-saturated liquid Th as calculated from the phase diagram. The $\overline{D}_N(ThN)$ values at temperatures of 1900 to 2400°C are illustrated in Fig. 1, p. 4, and show a possible downward curvature at the more elevated temperatures in the ln $\overline{D}_N$-vs.-1/T plot. Disregarding the possible curvature and assuming a straight line, the data are represented by the equation

$$\overline{D}_N(ThN) = 254 \cdot \exp - ((99400 \pm 10000)/RT) \text{ cm}^2/\text{s, temperature range } T = 2173 \text{ to } 2673 \text{ K}$$

where the uncertainty has been reduced to reflect the self consistency of the data and the activation energy is $E_A = 99.4 \pm 10.0$ kcal/g-atom N. The diffusion is suggested to occur by a vacancy mechanism in a nearly intact Th sublattice [30]. The chemical diffusion of N atoms in ThN is some 3 to 6 times faster than in the analogous UN crystals at the same temperature [31]. This is reasonable in view of the seemingly large unit cell dimension of ThN being about 5% greater than that of UN, and this difference persists up to 2400°C [32, 33]. The apparent activation energy for migration of the N atoms, however, is greater in the ThN than in the UN lattice.

Thermal Conductivity

Results of measurements on high-density arc-cast and zone-melted ThN at temperatures of 80 to 375 K by the guarded longitudinal heat flow technique of Moore et al. [34] have been reported by Weaver [35]. Thermal conductivity values for zone-melted ThN decrease linearly with increasing temperature from $\lambda = 67.74$ at 80.3 K to 57.52 $W \cdot m^{-1} \cdot K^{-1}$ at 135.6 K with the slope reported as $-0.261 \, W \cdot m^{-1} \cdot K^{-2}$. The slope changes near 140 K and there is a further linear but less steep decrease from $\lambda = 56.32$ at 151.8 K to 48.72 $W \cdot m^{-1} \cdot K^{-1}$ at 376.4 K with the slope given as $-0.031 \, W \cdot m^{-1} \cdot K^{-2}$. The data in the latter temperature range are represented to within 0.5% by the equation

$$\lambda \text{ (in } W \cdot m^{-1} \cdot K^{-1}) = 61.46 - 0.03388 \cdot T, \quad T = 150 \text{ to } 375 \text{ K}$$

from which the room temperature value is $\lambda_{298 K} = 51.36 \, W \cdot m^{-1} \cdot K^{-1}$, a value more than 6 times that of UO_2 and 3.7 times that of UN. For interpolated and extrapolated data for the temperature range 100 to 2000 K, see Table 6, p. 27 [35]. The Wiedemann-Franz-Lorenz ratio calculated with the above λ and R (equation (8), p. 26) data is nearly constant in the temperature range 80 to 300 K with the value $(\lambda \cdot R/T) = 1.2 \, L_o$, where $L_o = 2.443 \times 10^{-8} (V/K)^2$ is the Sommerfeld theoretical value for exclusive electronic transport. The observed coefficient is greater than 1 due to a lattice enhancement of the heat transport. Weaver resolved his data into its lattice, λ_l, and electronic, λ_e, components where $\lambda = \lambda_l + \lambda_e$, by the alloy analysis technique of Williams, Fulkerson [36] with the following results:

a) Electronic transport is the dominant mechanism of heat conduction in ThN over the entire temperature range investigated. There is a small electrical resistivity due to impurity which varies with temperature (deviation from the Matthiessen rule).

b) The Wiedemann-Franz-Lorenz ratio calculated with the electronic component, $L_e = \lambda_e R/T$, increases monotonically with temperature from 0.770 L_o at 80 K to 0.954 L_o at 300 K, approaching the L_o value for a degenerate electron gas.

c) The complementary lattice conductivity decreases monotonically from the value $\lambda_l = 22.1$ at 80 K to 9.88 $W \cdot m^{-1} \cdot K^{-1}$ at 300 K approximately as $\lambda \sim 43/T$, being governed mainly by phonon-phonon Umklapp scattering, but phonon-electron scattering is also significant.

d) Extrapolated heat conductivity data can be represented to within 1.7% by the equation λ (in $W \cdot m^{-1} \cdot K^{-1}) = 56.34 - 14.36 \times 10^{-3} T + 4.14 \times 10^{-6} T^2$ (T = 350 to 2000 K).

References for 4.1.3.3:

[1] Gerward, L., Olsen, J. S., Benedict, U., Itié, J.-P., Spirlet, J. C. (J. Appl. Cryst. **18** [1985] 339/41; also, Olsen, J. S., Steenstrup, S., Gerward, L., Benedict, U., Itié, J.-P., Rept. 8, Univ. Copenhagen Phys. Lab. II No. 85-14 [1985] 1/14; C.A. **103** [1985] No. 187192).

[2] Brooks, M. S. S. (J. Phys. F **14** [1984] 1157/71).

[3] Balankin, S. A., Danilkin, E. A., Loshmanov, L. P., Skorov, D. M., Sokolov, V. S. (Met. Metalloved. Chist. Metal. No. 13 [1979] 59/63).

[4] Benz, R., Hoffman, C. G., Rupert, G. N. (J. Am. Chem. Soc. **89** [1967] 191/7).

[5] Anonymous (PWAC-1019 [1965]).

[6] Ackermann, R. J., Rauh, E. G. (J. Chem. Thermodyn. **4** [1972] 521/32).

[7] Gingerich, K. A. (J. Chem. Phys. **49** [1968] 19/24).

[8] Rand, M. H. (At. Energy Rev. Spec. Issue No. 5 [1975] 7/85).

[9] Kubaschewski, O., Alcock, C. B. (Metallurgical Thermochemistry, 5th Ed., Pergamon, New York 1979).

[10] Aronson, S., Auskern, A. B. (J. Phys. Chem. **70** [1966] 3937/41).

[11] Kusakabe, T., Imoto, S. (Nippon Kinzoku Gakkaishi **35** [1971] 1115/20).

[12] Wagman, D. D., Evans, W. H., Parker, V. B., Schumm, R. H., Nuttall, R. L. (NBS-TN-270-8 [1981] 1/153; C.A. **69** [1982] No. 12232).

[13] Wagner, C. (Thermodynamics of Alloys, Addison-Wesley, Reading, Mass., 1952).

[14] Olson, W. M., Mulford, R. N. R. (J. Phys. Chem. **69** [1965] 1223/6).

[15] deNovion, C. H. (CEA-R-4113 [1970] 1/153; N.S.A. **25** [1971] No. 11980).

[16] deNovion, C. H., Costa, P. (Compt. Rend. B **270** [1970] 1415/8).

[17] Clusius, K., Franzosini, P. (Z. Naturforsch. **11a** [1956] 957).

[18] Decker, W. R., Finnemore, D. K. (Phys. Rev. [2] **172** [1968] 430/6).

[19] Gordon, J. E., Montgomery, H., Noer, R. J., Pickett, G. R., Tobón, R. (Phys. Rev. [2] **152** [1966] 432/7).

[20] Smith, P. L., Walcott, N. M. (Conf. Phys. Basses Temp., Paris 1955).

[21] Walcott, N. M., Hein, R. A. (Phil. Mag. [8] **3** [1958] 591/6).

[22] Danan, J. (CEA-R-4453 [1973]).

[23] Danan, J., deNovion, C. H., Dallaporta, H. (Solid State Commun. **10** [1972] 775/8).

[24] Dell, R. M., Martin, J. F. (At. Energy Rev. Spec. Issue No. 5 [1975] 7/85).

[25] Voitovich, R. F., Shakhanova, N. P. (Poroshkovaya Met. **1967** No. 3, pp. 75/9; Soviet Powder Met. Metal Ceram. **1967** No. 3, pp. 225/9).

[26] Tagawa, H. (Nippon Genshiryoku Gakkaishi **12** [1970] 658/65).

[27] Ono, F., Kanno, M., Mukaibo, T. (J. Nucl. Sci. Technol. [Tokyo] **10** [1973] 391/5).

[28] Aronson, S., Ingraham, A. (J. Nucl. Mater. **24** [1967] 74/9).

[29] Reimann, D. K., Kroeger, D. M., Lundy, T. S. (J. Nucl. Mater. **38** [1971] 191/6).

[30] Benz, R. (J. Electrochem. Soc. **119** [1972] 1596/602).

[31] Benz, R., Hutchinson, W. B. (J. Nucl. Mater. **36** [1970] 135/46).

[32] Benz, R., Balog, G. (High Temp. Sci. **3** [1971] 511/22).

[33] Benz, R., Balog, G., Baca, B. H. (High Temp. Sci. **2** [1970] 221/51).

[34] Moore, J. P., McElroy, D. L., Graves, R. S. (Can. J. Phys. **45** [1967] 3849/65).

[35] Weaver, S. C. (Diss. Univ. Tenn., Knoxville 1972).

[36] Williams, R. K., Fulkerson, W. (Proc. 8th Conf. Thermal Conductivity, West Lafayette, Ind., 1968 [1969], pp. 389/447).

4.1.3.4 Electrical, Magnetic, and Electronic Properties

Electrical Resistivity

The mononitride is an electronic conductor with a room-temperature specific resistivity first reported as $\approx 10^{-6}$ $\Omega \cdot m$ [15]. Electrical dc resistivity measurements (4 to 850 K) have been made on ThN specimens with porosities $P = 8$ to 13% that had been hot pressed at 1275°C and sintered at 1800°C in vacuum (10^{-5} Torr) for 6 h [16]. The results above 70 K, after correcting for porosity P with the formula $R = R_o (1-P)/(1+P/2)$, are represented by the equation

$$R \text{ (in } \mu\Omega \cdot m) = (7.6 + 0.608\,T) \times 10^{-3} \text{ in the temperature range 70 to 850 K} \quad \dots \quad (8)$$

The results below 70 K are fitted as a continuous curve matching the slope of equation (8) to the measured residual resistivity $R_{4.2\,K} = 0.03$ $\mu\Omega \cdot m$ with the Block-Grüneisen low-temperature limiting formula $(T/\Theta_D)^5$ and the Debye temperature of $\Theta_D = 300$ K [16]. Equation (8) gives the temperature coefficient $(1/R (300 \text{ K})) \cdot dR/dT = 3.20 \times 10^{-3} \text{K}^{-1}$ and the room-temperature value $R (300 \text{ K}) = 0.19$ $\mu\Omega \cdot m$. The latter value is lower by a factor of nearly 8 than the room-temperature value of 1.49 $\mu\Omega \cdot m$ reported for the analogous UN compound [17] but comparable to that of pure Th, viz. $R (300 \text{ K}) = 0.13$ to 0.164 $\mu\Omega \cdot m$ [18, 19] and $(1/R (300 \text{ K})) \cdot dR/dT = 3.5$ to $3.57 \times 10^{-3} \text{ K}^{-1}$ [19, 20].

References for 4.1.3.4 see pp. 31/2

Four-probe dc measurements later made on high-density zone-melted ThN over the temperature range 4 to 300 K by Weaver [21] are slightly lower in value but broadly confirm the above results of Auskern, Aronson [16]. The residual resistance of Weaver's [21] material was reported as being $R_{4.2K} = 0.021\ \mu\Omega \cdot m$. The results in the linear temperature region are reported as given by the equation R (in $\mu\Omega \cdot m$) = $(-21.22 + 0.63571\ T) \times 10^{-3} \pm 0.3\%$, temperature range 100 to 300 K with the room-temperature value of $R_{300K} = 0.1695\ \mu\Omega \cdot m$ and the temperature coefficient $(1/R\ (300\ K)) \cdot dR/dT = 3.75 \times 10^{-3}\ K^{-1}$. The following equation represents results of measurements (298 to 1300 K) by Kollie, according to Weaver [21]:

$$R\ (\text{in}\ \mu\Omega \cdot m) = (-3.47 + 0.58144\ T) \times 10^{-3},\ \text{temperature range 300 to 1700 K.}$$

From these results a formulation of the phonon component in the Block-Grüneisen formula leads to the Debye temperature of $\Theta_D = 532\ K$ which Weaver [21] notes is an unusually high value. Some typical R values are listed in Table 6.

Table 6

Interpolated Results of Measurements of the Specific Electrical Resistivity, R (4.2 to 1300 K), Thermal Conductivity λ (80.3 to 376.4 K), and the Absolute Thermoelectric Power, S_{ThN} (80.3 to 376.4 K) of High-Density Zone-Melted ThN [21]. (Extrapolated values are given in parentheses.)

temperature in K	R in $10^{-8} \cdot \Omega \cdot m$	λ in $W \cdot m^{-1} \cdot K^{-1}$	S_{ThN} in $\mu V/K$
4.2	2.1		
100	4.5	63.8	−1.0
200	10.5	54.3	+0.3
300	17.04	51.1	+0.4
400	22.9	48.0	−0.4
600	34.5	(48.4)	
800	46.2	(47.0)	
1000	57.8	(46.1)	
1200	69.4	(45.5)	
1400	81.1	(45.0)	
1600	(92.7)	(44.7)	
1800	(104)	(44.4)	
2000	(116)	(44.2)	

Thermoelectric Power (Seebeck Coefficient)

The Seebeck coefficient S of ThN was first stated to be a few $\mu V/K$ at room temperature [15]. Values relative to Cu from 4 to 375 K measured by the differential technique have been reported graphically by Auskern, Aronson [16]. From an extrapolation of the thermoelectric power data with the slope $-0.81 \times 10^{-8} (V/K)^2$ to 0 K they estimate the Fermi energy level from the equation $S = 2\pi^2 k^2/3e \cdot T/E_F$ as $E_F = 6\ eV$ and the electron effective mass as $0.7\ m_e$, where m_e is the electron rest mass.

Weaver [21] has reported results of measurements of the thermoelectric power, $P_{ThN,\ Constantan}$ obtained with arc-cast and zone-melted ThN in the temperature range 86 to 402 K. The data

References for 4.1.3.4 see pp. 31/2

converted to absolute values give $|S_{ThN}| \lesssim 1\ \mu V/K$ over the entire temperature range investigated; see Table 6, p. 27. Metallic Th also has a very small Seebeck coefficient, the value being $S_{Th} < 3\ \mu V/K$ up to 250 K [26, 27].

The thermoelectric figure of merit of ThN as calculated from Weaver's data, $Z = S^2/\lambda R = 3 \times 10^{-6}\ K^{-1}$ at 300 K, is an extremely low value compared, for example, to $Z = 2.2 \times 10^{-3}\ K^{-1}$ of the commercial material Ti_2Te_3 [28] and to $Z = 0.14 \times 10^{-3}\ K^{-1}$ of UN at 300 K [29]. Thus, ThN is not suited as a material for thermoelectric conversion.

Superconductivity

A transition to the superconducting state occurs at $T_c = 3.2$ K [30] and 3.16 K [31]. The T_c decreases linearly with increasing hydrostatic pressure up to 26.7 kbar with the pressure coefficient $dT_c/dp = -14.6 \pm 0.3$ mK/kbar $= -14.6 \times 10^{-2}$ mK/MPa [31]. In comparison, pure Th metal becomes superconducting at $T_c = 1.37$ [32, 33], 1.39 [6], and 1.4 K [34] and has the pressure coefficient $dT_c/dp = -14.7 \pm 0.2$ mK/kbar $= -14.7 \times 10^{-2}$ mK/MPa, which is constant up to 20 kbar = 2000 MPa [35].

Hall Coefficient

Auskern, Aronson [16] have made measurements of the Hall coefficient R_H on two hot-pressed ThN specimens with porosities P = 8 and 13% in magnetic fields up to 13 kG. The result, corrected for porosity with the Maxwell equation $R_H = H_{Ho}(1 - P)$, is $R_H = -1.6 \times 10^{-14}$ V·m·A^{-1}·G^{-1} = -1.6×10^{-4} cm^3/C independent of temperature from 4 to 375 K [16]. The number of charge carriers calculated on the assumption of a single conduction band [36] and only electrons as charge carriers is $n = 1/e \cdot R_H = -3.9 \times 10^{22}$ cm^{-3} = 1.34 electrons/Th atom. The intrinsic mobility is $\mu = 8.4$ cm^2·V^{-1}·s^{-1} [37]. The R_H of ThN is very nearly the same as that of Th metal which has been reported for two specimens as $R_H = -1.3 \times 10^{-14} = -1.3 \times 10^{-4}$ and -1.1×10^{-14} V·m·A^{-1}·G^{-1} = -1.1×10^{-4} cm^3/C in magnetic fields of 3.7 to 4.5 kG [38, 39]. In the same approximation, $n = 4.7 \times 10^{22}$ cm^{-3} = 1.5 electrons per Th atom in the metal at room temperature.

Thermionic Emission

For practical comparison of the suitability of different kinds of refractories as cathode materials in generating high electronic emission currents, a figure of merit was defined by the equation
$$F = 2 \log T_m - 5 \times 10^3\ \Phi/T_m,$$
where T_m is the melting point (in K) and Φ (in eV) is the work function. This definition is based upon the Richardson-Dushman equation for thermionic emission. Practically, F provides a suitable measure of the combined thermionic and field emission currents. On this basis, the fcc ThN is rated as a moderately good electron emitter with a calculated $F = +1.6$ (read from a plot) value that is higher than that of high-melting metals, e.g., Ta and W, but lower that calculated for the thermionic oxides such as CaO, ThO$_2$, etc. and a number of carbides such as TiC, TaC, CeC$_2$, and others [42].

Magnetic Susceptibility

Crystals of ThN were at first stated to be diamagnetic [1] but in the first reported quantitative measurements by Aronson, Auskern [2], the susceptibility was given as $\chi = +0.169$ to 0.184×10^{-6} (41.6 to 45.3×10^{-6}) at 83 K, 0.166 to 0.183×10^{-6} (40.8 to 45.0×10^{-6}) at 197 K,

References for 4.1.3.4 see pp. 31/2

and 0.165 to 0.178×10^{-6} cm³/g (40.6 to 43.8×10^{-6} cm³/mol) at 295 K with only a slight negative slope in its variation with temperature. Raphael, deNovion [3] later reported χ to be independent of temperature over the broader range of 4 to 1000 K with the value 0.142×10^{-6} cm³/g $= 35 \times 10^{-6}$ cm³/mol, in essential agreement with Aronson, Auskern [2]. Adachi, Imoto [4] give the value 0.22×10^{-6} cm³/g $= 54 \times 10^{-6}$ cm³/mol as constant over the temperature range of 77 to 300 K. These results indicate the absence of any localized or permanent moment on the Th atom in ThN, the observed paramagnetism being due to conducting electrons [2]. There is no orbital moment from the 6d or spin moment from the 5f subshell, the moments being quenched in formation of a delocalized conducting band.

The element N is diamagnetic. The paramagnetism of metallic Th is the sum of contributions from the ion core diamagnetism, a (probably negligible) localized positive Van Vleck paramagnetism, and Pauli spin paramagnetism of the collective free electrons in the conducting band with the net value $\chi = +0.412 \times 10^{-6}$ cm³/g (300 K) [5]. It has a small temperature coefficient up to 300 K [6]. The paramagnetism of ThN is similarly derived from that of the free electrons χ, being the net sum of a diamagnetic (core electrons) and a nearly temperature-independent Pauli spin paramagnetism contributed by the conducting electrons [2 to 4, 7]. With corrections for diamagnetism on this basis, the net paramagnetic component of ThN has been estimated as 70×10^{-6} [3] and 75×10^{-6} cm³/mol [8]. Spin paramagnetism as calculated from the electronic specific heat data is in reasonable agreement with the observed magnetic susceptibility indicating that any contributions from the diamagnetic or orbital paramagnetic susceptibilities are small. Allbutt, Dell [9] state that their unpublished susceptibility measurements support the existence of the Th^{4+} ion state in ThN.

Nuclear Magnetic Resonance, Knight Shift

Results of continuous wave NMR measurements have been reported for ^{14}N [10] and for ^{15}N [11 to 13] in ThN. Kuzamitz [10] reported symmetrical peaks with a line width of 22 ± 5 Oe independent of temperature from 77 to 300 K and of applied magnetic field, with the width being attributed to quadrupole interactions with the ion electric field gradients. Boutard et al. [13] reported line widths of 5 ± 1 Oe to be independent of temperature (4.2 to 300 K) and of field (6.5 to 10 kOe). Knight shifts relative to NH_4^+, $K = B_0(x) - B_0(ref)/B_0(ref)$, where $B_0(x)$ and $B_0(ref)$ are the magnetic fields at resonance of the nuclei in the metallic substance and in the diamagnetic reference material, respectively, have been reported with the values of $K = +(10.7 \pm 1.5) \times 10^{-4}$ for ^{14}N [10] and as $+(8.8 \pm 0.3) \times 10^{-4}$ for ^{15}N in ThN [13]. The appreciable Knight shift confirms the existence of conducting electrons. Kuznietz [10] considers his results to be consistent with a ThN lattice built up of diamagnetic Th^{4+} and N^{3-} ions with one free electron for each Th atom. The product of temperature and spin-lattice relaxation time is nearly independent of temperature, being $TT_1 = 882 \pm 63$ s·K at 4.2 K and 785 ± 77 s·K at 77 K. The Korringa constant enhancement factor $\Delta = (K^2TT_1)_{exp}/(K^2TT_1)_{theory} = 23.8 \pm 2.7$ at 77 K is due largely to orbital contributions to the Knight shift [11, 13].

Mössbauer Spectroscopy

The Mössbauer spectrum from the first excited state of ^{232}Th in a Th metal target and resonance absorbed at 78 K in ThN, as well as that with other combinations of Th and ThC targets and ThO_2, Th, and ThC_2 absorbers at 25 and 30 K, were determined by resonance absorption following Coulomb excitation by 6.0-MeV He^{2+} (RACE method) [14]. The recoil-less fractions $f_a = 0.35 \pm 0.04$ for the absorber and $f_s = 0.31 \pm 0.06$ for the source were deduced corresponding to the respective Debye temperatures $\Theta_D = 121 \pm 13$ and 121 ± 20 K. The zero-thickness natural absorber line of the spectrum was found to be $\Gamma_0 = 13.0 \pm 1.3$ mm/s with ThN as compared to 11.35 ± 0.26 mm/s with ThO_2. The latter value, after correction for broadening

References for 4.1.3.4 see pp. 31/2

due to the experimental geometry, gives the absorber line width of 9.6 ± 0.4 mm/s for the first excited state of ^{232}T. An isomer shift for the metal target vs. ThO_2 absorber of $+0.8 \pm 0.09$ mm/s was measured.

Photoelectron Spectra

The X-ray photoelectron spectrum (XPS) obtained with $h\nu = 1486.6$ eV(AlKα) radiation and the ultraviolet photoelectron spectrum (UPS) obtained with $h\nu = 40.8$ eV radiation has been reported for ThN and compared with that of "clean" Th by Norton et al. [40]. As illustrated in **Fig. 4**a, the ultraviolet spectrum of Th has a peak associated with 6d levels within 1.3 eV of the Fermi level. *In situ* nitridation of the Th metal results in the complete removal of this peak and the appearance of an emission peak 4 eV higher in binding energy in the nitride spectrum, Fig. 4a. This latter peak is associated with 6d levels stabilized in ThN by 4 eV relative to that in Th. The implication is that the d electrons calculated to be 2.6 ± 0.7 electrons/Th atom in number in the metal participate in chemical bonding of ThN mostly with Th 6d-N2p character. Veal, Lam [41] have compared the ultraviolet spectrum of the analogous UN crystals with that of ThN and note an additional strong multiplet of peaks in a narrow energy range near $E_F = 0$ in the spectrum of UN. They attribute the latter multiplet to 5f electrons.

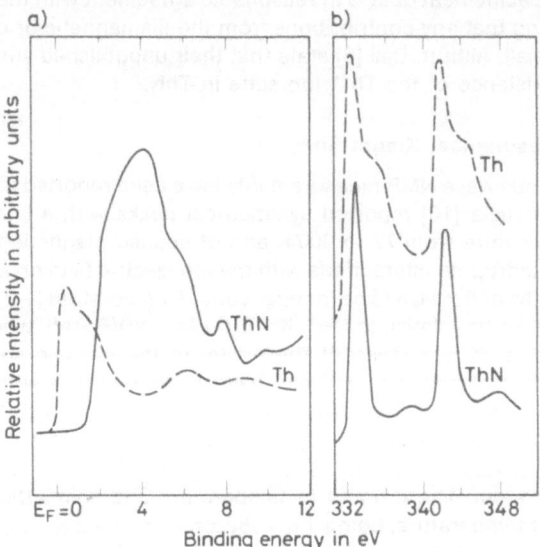

Fig. 4
Ultraviolet ($h\nu = 40.8$ eV) and X-ray ($h\nu(AlK_\alpha) = 1486.6$ eV) photoemission spectra of ThN and Th [40].

The XPS data as illustrated in Fig. 4b for Th and ThN show two principal 4f peaks in both spectra that are very symmetrical and have line widths of 1.6 eV. The peaks in the Th spectrum have shoulders displaced by 2.5 eV on the high binding energy side. These shoulders disappear on conversion to ThN. This is attributed to a shake-up transition involving excitation of 6d electrons of Th. Corresponding emission peaks appear in the ThN spectrum but displaced by $\Delta E = 5.5$ eV to higher binding energies as compared to the expected 4 eV [40].

Band Structure

The band structure of metallic Th and ThN has been calculated by Imoto et al. [43] using the "tight-binding" method with parametrization refinement. Values of 2-center parameters and the spin-orbit coupling parameter were chosen for Th so that the orthogonal plane waves describing the outer s electrons fit the relativistic plane-wave calculations of Gupta, Loucks [44, 45] with the f levels deleted. The results were applied to ThN. The thus calculated density of states at the Fermi level $N(E_F) = 0.99$ eV^{-1}·atom^{-1} can be compared with the calorimetrically measured value of 1.32 eV^{-1}·atom^{-1}. Calculated electron populations show that the occupied states in ThN are mainly of the σ-type bonding between N p and the Th dγ and of the π-type bonding between the Np and Th dε orbitals. The bonding in ThN is thus concluded to be due principally to interactions between p and d states.

The volume contribution by the metallic 5f-5f and the covalent cation 5f-N2p bonds to a virial-theorem formulation of the equations of state of a series of light actinide nitrides was calculated in the self-consistent linear muffin tin orbital (LMTO), relativistic LMTO, and spin-polarized LMTO approximations [46]. The results for ThN give the same lattice spacing in all three approximations higher by ca. 3% than the experimental value, which discrepancy is attributed to the assumed frozen core ions [47].

References for 4.1.3.4:

[1] Didchenko, R., Gortsema, F. P. (Inorg. Chem. **2** [1963] 1079/80).
[2] Aronson, S., Auskern, A. B. (J. Chem. Phys. **48** [1968] 1760/5).
[3] Raphael, G., deNovion, C. (Solid State Commun. **7** [1969] 791/3).
[4] Adachi, H., Imoto, S. (Technol. Rept. Osaka Univ. **23** [1973] 1121/54, (Eng.) 425/9).
[5] Freemann, A. J., Koelling, D. D. (in: Freemann, A. J., Darby, J. B., The Actinides, Vol. 1, Electronic Energy Band Structure of the Actinide Metals, Academic, New York 1974, pp. 51/108).
[6] Greiner, J. D., Smith, J. F. (Phys. Rev. [3] B **4** [1971] 3275/7).
[7] Dell, R. M., Bridger, N. J. (in: Bagnall, K. W., Lanthanides and Actinides, Actinide Chalcogenides and Pnictides, Ser. 1, Vol. 7, Butterworth, London 1972, pp. 211/74).
[8] deNovion, C. H. (CEA-R-4113 [1970] 1/153; N.S.A. **25** [1971] No. 11 980).
[9] Allbutt, M., Dell, R. M. (J. Inorg. Nucl. Chem. **30** [1968] 705/10).
[10] Kuznietz, M. (J. Chem. Phys. **49** [1968] 3731/2; Phys. Rev. [2] **180** [1969] 476/81).

[11] Boutard, J. L. (CEA-R-4797 [1976] 1/188; C.A. **87** [1977] No. 60365).
[12] Boutard, J. L., deNovion, C. H. (Plutonium 1975 Other Actinides Proc. 5th Intern. Conf., Baden-Baden, FRG, 1975 [1976], pp. 505/13).
[13] Boutard, J. L., deNovion, C. H., Alloul, H. (J. Phys. [Paris] **41** [1980] 845/53).
[14] Durkee, P., Hershkowitz, N. (KFK-2098 [1974]; Phys. Rev. [3] B **3** [1971] 3607/15).
[15] Didchenko, R., Gortsema, F. P. (Inorg. Chem. **2** [1963] 1079/80).
[16] Auskern, A. B., Aronson, S. (J. Phys. Chem. Solids **28** [1967] 1069/71).
[17] Fulkerson, W., Kollie, T. G., Weaver, S. C., Moore, J. P., Williams, R. K. (Plutonium 1970 Other Actinides Proc. Intern. Conf., 1970, pp. 374/85).
[18] Smith, J. F. (Metal Thorium Proc. Conf., Cleveland 1956 [1958], pp. 133/47).
[19] Peterson, D. T., Page, D. F., Rump, R. B., Finnemore, D. K. (Phys. Rev. [2] **153** [1967] 701/4).
[20] Cuthbert, F. L. (Thorium Production Technology, Chapter 6, Preparation of Th Metal by Reduction, Addison-Wesley, Reading, Mass., 1958, pp. 146/201).

[21] Weaver, S. C. (Diss. Univ. Tenn., Knoxville 1972).

[22] Borelius, G., Kesson, W. H., Johansson, C. H., Linde, J. O. (Proc. Koninkl. Ned. Akad. Wetenschap. **35** [1932] 10/4).

[23] Christian, J. W., Jan, J. P., Pearson, W. B., Templeton, I. M. (Proc. Roy. Soc. [London] A **245** [1958] 213/21).

[24] Cusack, N., Kendall, P. (Proc. Phys. Soc. [London] **72** [1958] 898/901).

[25] Barnard, R. D. (Thermoelectricity in Metalls and Alloys, Taylor & Francis, London 1972).

[26] Haen, P., Meaden, G. T. (Cryogenics **5** [1965] 194/8).

[27] Meaden, G. T. (Proc. Roy. Soc. [London] A **276** [1963] 553/7).

[28] Rosi, F. E., Ramberg, E. G. (in: Egli, P. H., Thermoelectricity, Wiley, New York, 1960, pp. 121/55).

[29] Moser, J. B., Kruger, O. L., Handwerk, J. H. (Proc. Brit. Ceram. Soc. No. 10 [1968] 129/39).

[30] Giorgi, A. L., Szklarz, E. G., Krupka, M. C. (LADC-13299 [1970] 1/7; N.S.A. **26** [1972] No. 21559).

[31] Dietrich, M. (KFK-2098 [1974]).

[32] Wolcott, N. M., Hein, R. A. (Phil Mag. [8] **3** [1958] 591/6).

[33] Gordon, J. E., Montgomery, H., Noer, R. J. Pickett, G. R., Tobón, R. (Phys. Rev. [2] **152** [1966] 432/7).

[34] Fowler, R. D., Matthias, B. T., Asprey, L. B., Hill, H. H., Lindsay, J. D. G., Olsen, C. E., White, R. W. (Phys. Rev. Letters **15** [1965] 860/1).

[35] Fertig, W. A., Moodenbaugh, A. R., Maple, M. B. (Phys. Letters A **38** [1972] 517/8).

[36] Ziman, J. M. (Electrons and Phonons, Oxford Clarendon Press, London 1960, p. 495).

[37] Auskern, A. B., Aronson, S. (J. Appl. Phys. **41** [1970] 227/32).

[38] Berlincourt, T. G. (Phys. Rev. [2] **114** [1959] 969/77).

[39] Bodine Jr., J. H. (Phys. Rev. [2] **102** [1956] 1459).

[40] Norton, P. R., Tapping, R. L., Creber, D. K., Buyers, W. J. L. (Phys. Rev. [3] B **21** [1980] 2572/7).

[41] Veal, B. W., Lam, D. J. (CONF-790917-17 [1979] 1/20; INIS Atomindex **11** [1980] No. 52150).

[42] Zaima, S., Adachi, H., Shibata, Y. (J. Vac. Sci. Technol [2] B **2** [1984] 73/8).

[43] Imoto, S., Adachi, H., Hori, T. (J. Nucl. Sci. Technol. **12** [1975] 711/6).

[44] Gupta, R. P., Loucks, T. L. (Phys. Rev. Letters **22** [1969] 458/61).

[45] Gupta, R. P., Loucks, T. L. (Phys. Rev. [3] B **3** [1971] 1834/42).

[46] Brooks, M. S. S. (J. Magn. Magn. Mater. **29** [1982] 257/61; J. Phys. F **14** [1984] 857/71).

[47] Brooks, M. S. S. (J. Phys. F **14** [1984] 1157/71).

4.1.3.5 Optical Properties

Color. ThN is golden yellow [1, 2].

Emissivity. The spectral emissivity, $E_{0.65}$, at the wave length $0.65\,\mu m$ is defined by the equation

$$(1/T)-(1/T_s) = (0.65\ \mu m/14350\ \mu m \cdot K)\ln E_{0.65},$$

where T_s is the observed surface temperature. Measurements of $E_{0.65}$ as a function of T have been made with a disc-shaped specimen presumably formed by cold pressing of ThN powders

References for 4.1.3.5 see pp. 34

and heating in vacuum on a W strip [3, 4]. The reported results are listed in Table 7 together with calculated T_s values. These data should be used with reservation because of imcomplete description of the experimental conditions. In an open vacuum system, ThN can decompose more or less rapidly to Th metal and the emissivity can be expected to change. The emissivity values given in Table 7 are appreciably higher than the value $E_{0.667} \mu m = 0.380$ (1300 to 1700 K) reported for solid [5] and $E_{0.65} \mu m = 0.40$ for liquid Th [6].

Table 7

Spectral Emissivity of ThN, $E_{0.65}$, with the Wave Length $\lambda = 0.65$, μm as a Function of Temperature. T_s Denotes the Surface Temperature [3, 4].

T in K	1781	1839	1898	1953	2013	2188
T_s in K	1766	1817	1865	1913	1963	2113
$E_{0.65}$	0.900	0.865	0.814	0.790	0.756	0.699

Infrared Absorption Spectra. Molecular spectra of matrix-isolated $Th^{14}N$ and $Th^{15}N$ at 15 K as produced by sputtering Th in $Ar + N_2$ and in $Kr + N_2$ mixtures have been reported by Green, Reedy [7]. The infrared spectra show a multiplet of absorption peaks (Table 8) all corresponding to one and the same stretching mode of vibration but to several different matrix sites characterized by different arrangements of the inert gas atoms. The site for $Th^{14}N$ with the absorption peak $\bar{\nu} = 934.61 \, cm^{-1}$ predominates after the matrix is annealed at 34 K.

Table 8

Infrared Absorption Spectra of Matrix-Isolated ThN in Solid Ar and in Solid Kr at 14 K. The Peaks are Assigned to the Given Harmonic and Anharmonic Vibrational Constants of ThN [7].

$\bar{\nu}$ in $cm^{-1} \pm 0.05$		vibrational constants	
$Th^{14}N$	$Th^{15}N$	ω_e in $cm^{-1} \pm 1.0$	$\omega_e x_e$ in $cm^{-1} \pm 0.5$
Argon			
934.61	905.07	941.9	3.7
924.75	895.50	941.3	3.3
923.11	893.92	929.9	3.4
Krypton			
927.51	898.25	936.5	4.5
917.38	888.39	924.7	3.7
915.21	886.26	921.6	3.2
913.9	885.0		

Spectra obtained with higher N_2 concentrations reveal a $Th-N_2$ complex (not a dinitride) having a "sideways" bonding of N_2 to Th with C_{2v} symmetry and equivalent N atoms. The N–N stretching frequencies $\bar{\nu}_{N-N} = 1828.59 \, cm^{-1}$ in $Th^{14}N$, 1798.47 in $Th(^{14}N, ^{15}N)$, and 1767.78 cm^{-1} in $Th^{15}N_2$ indicate a strong $Th-N_2$ interaction and an ionic $Th^-N_2^+$ contribution to the bonding.

References for 4.1.3.5 see p. 34

References for 4.1.3.5:

[1] Chiotti, P. (AECD-3072 [1951] 1/63; N.S.A. **5** [1951] 3141).
[2] Gerds, A. F., Mallett, M. W. (J. Electrochem. Soc. **101** [1954] 175/80).
[3] Kusakabe, T., Imoto, S. (Nippon Kinzoku Gakkaishi **35** [1971] 1115/20).
[4] Kusakabe, T., Imoto, S. (Technol. Rept. Osaka Univ. **22** [1972] 477/88).
[5] Whitney, L. V. (Phys. Rev. [2] **48** [1935] 458/61).
[6] Roesner, Wensel (in: Weast, R. C., Handbook of Chemistry and Physics, 51th Ed., The Chemical Rubber Co., Cleveland, Oh., 1970 [1971], p. E-235).
[7] Green, D. W., Reedy, G. T. (J. Mol. Spectrosc. **74** [1979] 423/34).

4.1.3.6   Chemical Behavior

For thermal decomposition, see p. 20.

Reaction with O_2. The rate of reaction of ThN powder (mesh 150 to 400) with pure O_2 has been measured gravimetrically. The rate of the reaction

$$ThN + O_2 \rightarrow ThO_2 + \tfrac{1}{2}N_2$$

is independent of pressure (0.2 to 1 atm) in the temperature range 360 to 420°C and is controlled by a surface mechanism. After a substantial extent of reaction, the time course of reaction follows the equation $1-(1-c)^{\frac{1}{3}} = kt$ with the activation energy of 16 ± 2 kcal/mol, where c denotes the conversion ratio at time t and k is a constant. The reaction is too slow to follow below 360°C and is nearly complete in 6 h at 480°C. ThN powder ignites at 520°C and above [1].

Reaction with Air. Reaction takes place in moist air to form at first a dark and then a white ThO_2 powder often visible within an hour. The oxidation is slower in dry air but to avoid oxygen contamination storage and manipulation are done in a purified inert or N_2 atmosphere. Because of the sensitivity to oxidation even by trace amounts of water $+ O_2$, the nitrides are not practical as catalysts.

Reaction with Graphite. The rate of reaction of ThN with C in loose powder mixtures in the mole ratio C/Th = 1 has been measured gravimetrically by Ozaki et al [2]. Three different particle sizes were investigated, coarse (150 to 65 mesh), medium (400 to 150 mesh), and fine (−400 mesh) with closely corresponding graphite particle sizes. The value of x in the formula $ThN_{1-x}C_x$ was determined by X-ray diffractometry. Under the experimental conditions, the value of x increases linearly with the extent of the reaction

$$ThN + C = ThN_{1-x}C_x + x/2\,N_2 + (1-x)\,C \quad (1420 \text{ to } 1630°C)$$

given as the weight change ratio, α, denoting the fraction of the total N displaced. Although the data show that there was some possible mechanical loss of C in the experiments, the kinetics of reaction in the initial stages of time, t, follows the equation

$$-\log (1-\alpha) = kt,$$

indicating that the dominating mechanism is similar for all 3 particle sizes. The reaction rate constant, k, is greater for the fine particle size than for the medium and coarse, both of which exhibit nearly the same k values. The energy of activation, $E_A = 62 \pm 3$ kcal/mol, is essentially independent of particle size.

Reaction with U. The principal product of reaction of Th metal with UN at 1000°C is the bright yellow ThN, indicating the reaction is $UN + Th \rightarrow ThN + U$ [3]. Apart from small mutual solubilities between the phases, this shows that the reverse of this reaction does not occur spontaneously, i.e., ThN does not react with U, in agreement with the available thermodynamic data on the formation of UN and ThN under standard conditions [4].

Reaction with H₂O. The powder reacts with liquid water evolving H_2 gas by the reaction $ThN + 2H_2O \rightarrow ThO_2 + NH_3 + \frac{1}{2}H_2$. The reaction rate is given roughly by an equation of the form $\log(c_\infty/(c_\infty - c)) = kt$ (5 to 70°C), where c is the NH_4^+ ion concentration at time t and c_∞ is that upon completion of the reaction. The apparent activation energy is 15 kcal/mol (5 to 70°C). A surface reaction mechanism is rate-controlling. In contrast to the analogous UN compound, which produces an appreciable amount of a higher nitride phase as an intermediate, the hydrolysis of ThN does not yield a higher nitride. Crystals of ThN dissolve rapidly in moderately concentrated strong acids [5].

Reaction of ThN with water vapor gives the same products as with liquid water and is 90% complete in 20 min at 200°C [5].

References for 4.1.3.6:

[1] Ozaki, S., Kanno, M., Mukaibo, T. (J. Nucl. Sci. Technol. [Tokyo] **8** [1971] 41/4).
[2] Ozaki, S., Ono, Y., Kanno, M. (J. Nucl. Sci. Technol. [Tokyo] **10** [1973] 374/8).
[3] Katz, S. (J. Nucl. Mater. **6** [1962] 172/81).
[4] Kubaschewski, O., Alcock, C. B. (Metallurgical Thermochemistry, 5th Ed., Pergamon, New York 1979).
[5] Sugihara, S., Imoto, S. (J. Nucl. Sci. Technol. [Tokyo] **8** [1971] 630/6).

4.1.4 Trithorium Tetranitride Th₃N₄

4.1.4.1 Preparation

Nitridation of Thorium or Thorium Hydride. The reactions of Th or ThH_4 with N_2 as described in Section 4.1.3.1, p. 11, are most commonly used for preparation whereby Th_3N_4 is obtained as an intermediate.

Reaction of K₂Th(NO₃)₆ with KNH₂. Liquid NH_3 solutions of anhydrous $K_2Th(NO_3)_6$ and $4KNH_2$ are mixed and the resulting $Th_2N_2(NH)$ precipitate, after isolation, is heated in a stream of N_2 to 130°C, where it decomposes to a lemon yellow amorphous Th_3N_4 [1].

ThCl₄ Nitridation. Th_3N_4 is deposited in small amounts on a tungsten wire electrically heated to 1000°C or above in a gas mixture of N_2 and either $ThCl_4$ or $H_2 + ThCl_4$ vapors [2]. It is not clear whether or not exchange of Cl with W is involved.

Preparation of Metastable β-Th₃N₄. This compound is obtained from $Th_2N_2(NH)$ [3]. The latter compound is formed by reacting Th metal powder with NH_3 in a pressure vessel, cf. p. 60. Subsequent thermal decomposition of the $Th_2N_2(NH)$ at 700°C until gas evolution is practically complete or in vacuum at 100 to 390°C gives the brown β-Th_3N_4 compound, which is identified by its characteristic X-ray diffraction pattern. The decomposition takes place with a continuous shift of the diffraction peaks indicating an extensive mutual solubility between the $Th_2N_2(NH)$ and β-Th_3N_4 phases. The composition of the β-Th_3N_4 is based upon chemical analyses for Th and N after allowance is made for 4.95% ThO$_2$ in the starting metal. The β-Th$_3$N$_4$ compound is also observed as a component in Th$_2$N$_2$O + Th$_3$N$_4$ phase mixtures quenched from 1200 to 1850°C [3] and as a minor component in C-saturated Th–C–N phase mixtures at 1300 to 1500°C [4] but phase purity was not realized under the latter conditions.

References for 4.1.4.1:

[1] Schmitz-Dumont, O., Raabe, R. (Z. Anorg. Allgem. Chem. **277** [1954] 297/314).
[2] Düsing, W., Hüniger, M. (Tech. Wiss. Abhandl. Osram-Ges. **2** [1931] 357/65).
[3] Juza, R., Gerke, H. (Z. Anorg. Allgem. Chem. **363** [1968] 245/57).
[4] Pialoux, A. (J. Nucl. Mater. **91** [1980] 127/46).

4.1.4.2   Physical Properties

Crystal Structure, Density

The trigonal, Al$_4$C$_3$-type, crystal structure has been established from X-ray powder diffraction data [1] and the assumed N atom positions were later essentially confirmed by neutron diffractometry [2]. Reported values for the hexagonal lattice parameters are summarized in Table 9. The value a = 387.5 ± 0.2 pm given in row 5 of Table 9 is based upon the neutron diffraction data and is appreciably larger than the value given in row 1 for material prepared in exactly the same way but measured by X-ray powder diffraction film technique. The accuracy of the values derived from the neutron diffraction results is clearly slightly more limited than that from the X-ray technique, evidently because of an insufficient number of well resolved neutron peaks at high angles. The Th$_3$N$_4$ crystal structure is as follows:

Space group R$\bar{3}$m-D$_{3d}^5$ (No. 166) [3].

Hexagonal description: a_H = 387.1 ± 0.1 pm, c_H = 2738.5 ± 0.5 pm.

Rhombohedral description: a_R = 939.8 ± 0.2 pm, α = 23.78° ± 0.01°.

Atomic positions (Th based on X-ray [1] and N atoms based on neutron diffraction data [2]):

hexagonal description	rhombohedral description
(0,0,0)(1/3, 2/3, 1/3)(2/3, 1/3, 2/3) +	
3Th(I) in (0,0,0);	1Th(I) in (0,0,0);
6Th(II) in ± (0,0,z_{Th});	2Th(II) in ± (z_{Th}, z_{Th}, z_{Th});
6N(I) in ± (0,0,$z_{N(I)}$);	2N(I) in ± ($z_{N(I)}$, $z_{N(I)}$, $z_{N(I)}$);
6N(II) in ± (0,0,$z_{N(II)}$);	2N(II) in ± ($z_{N(II)}$, $z_{N(II)}$, $z_{(II)}$);

where z_{Th} = 0.2221 ± 0.0004, $z_{N(I)}$ = 0.1320 ± 0.004, and $z_{N(II)}$ = 0.3766 ± 0.0004. The theoretical density is D_X = 10.54 ± 0.01 g/cm^3.

The atomic configuration in the hexagonal unit cell is illustrated in **Fig. 5**. The Th atoms have the close-packed configuration ABABCBCACA ... observed in the samarium metal structure and the c axis is 9 times the separation of the hexagonal layers. The N(I) atoms

References for 4.1.4.2 see pp. 41/2

occupy tetrahedral and the N(II) occupy octahedral holes of the close-packed metal configuration. The following interatomic distances (in pm) have been derived from the lattice parameters as based upon the X-ray diffraction data and the N atomic position parameters based upon the neutron diffraction data [1, 2]:

Th–Th		Th–N	N–Th
Th(II)–3Th(II)	377 ± 2	Th(II)–3N(I) 231 ± 1	N(I)–3Th(II) 231
Th(I)–6Th(I)	387.1 ± 0.1	Th(II)–1N(I) 247 ± 2	N(I)–1Th(II) 247
Th(I)–6Th(II)	378 ± 1	Th(I) –6N(II) 253 ± 1	N(II)–3Th(I) 253
Th(II)–3Th(I)	378 ± 1		N(II)–3Th(II) 291 ± 1
Th(II)–6Th(II)	387.1 ± 0.1		

The mean bond distances in this ionic compound are Th(II)–4N(I) = 235 and Th(I)–6N(II) = 253 pm. The latter value is even larger than the bond distance Th–8O = 242.4 pm of Th with the coordination number of 8 in ThO_2 showing that the N^{3-} ion radius is larger than that of the O^{2-} ion by about 17 pm in these compounds, allowing for the difference in coordination [11].

Table 9

Reported Hexagonal Lattice Parameters of Trigonal Th_3N_4 at Room Temperature.

a_H in pm	c_H in pm	D_X in g/cm³	Ref.
387.1 ± 0.1	2738.5 ± 0.5	10.54 ± 0.01	[1]
386.9 ± 0.2	2740	10.55 ± 0.02	[4]
386.9 ± 0.1	2741 ± 3	10.55 ± 0.02	[5]
386.6	2736.8	10.57	[6]
387.5 ± 0.2	2739 ± 4	10.52 ± 0.03	[2]
387.4 ± 0.3	2739.2 ± 0.6	10.52 ± 0.02	[8]
386.6	2740.9	10.56	[9, 10, 26]

The Th_3N_4 structure is closely related to that of Th_2N_2O the unit cell of which is illustrated in Fig. 21, p. 67. The configuration about the Th(II) atoms of the Th_3N_4 structure is almost identical with that of Th in Th_2N_2O; however, the Th(I) atoms form 6 bonds to N atoms, all occupying octahedral holes [1].

A hypothetical process (considerations of the authors of this article) by which the Th_2N_2O can be visualized as transforming into the Th_3N_4 structure is as follows: replace the O in each Th_2N_2O unit by N atoms and stack the resulting Th_2N_3 layers with a hexagonal Th A̲ layer inserted between every 4th layer to give the correct chemical composition, viz. $4Th_2N_3 + Th = 3Th_3N_4$, the correct N configuration, and the Th layer sequence ... ABCBCBCBCA ... With the N atoms held fixed the 3rd Th(C̲) layer is subjected to the glide $G = (\frac{1}{3}\vec{a}_1 + \frac{2}{3}\vec{a}_2)$ and the 8th (B) layer to the inverse glide, $\overline{G}$. Then a slight contraction of the resulting lattice and small rearrangement of the N atoms gives the Th_3N_4 structure. Other compounds with related structures are $Th_3N_2(NH)_3$, $Th(NH)_2$, $ThN(NH_2)$, β-Th_3N_4, and $Th_2N_2(NH)$.

References for 4.1.4.2 see pp. 41/2

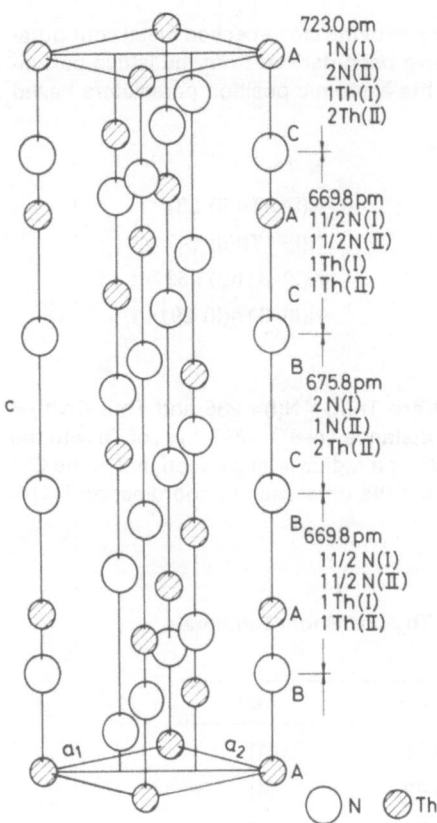

723.0 pm
1N(I)
2N(II)
1Th(I)
2Th(II)

C

669.8 pm
11/2 N(I)
11/2 N(II)
1Th(I)
1Th(II)

C

B
675.8 pm
2N(I)
1N(II)
2Th(II)

C

B
669.8 pm
11/2 N(I)
11/2 N(II)
1Th(I)
1Th(II)

c

a₁   a₂   A

B

A

○ N   ◍ Th

Fig. 5
Atomic configuration in the hexagonal unit cell of trigonal Th₃N₄, drawn by the authors of this article, showing the different stacking of the hexagonal layers denoted by A, B, and C, and the Th₃N₄ units which are compared in the text with Th₂N₂O.

Metastable β-Th₃N₄. Crystals of this compound have a monoclinic deformed Th_2N_2O structure with the lattice parameters a = 695.2, b = 383.0, c = 620.6 pm and β = 90.71°. The reported pycnometric density is D(25°C) = 10.13 g/cm³ as compared to the theoretical density of 10.08 g/cm³ calculated by Juza, Gerke [6] for the unit cell content of 4($ThN_{4/3}$) = $Th_4N_{16/3}$. The suggested structure with the space group $Bm-C_s^3$ (No. 8) has a monoclinically deformed hexagonal close-packed Th lattice and is related to the ternary compound $Th_2N_2(NH)$ [6].

Lattice Thermal Expansion

High-temperature hexagonal lattice parameters of Th₃N₄ have been measured at temperatures up to 1905°C. The data show an appreciable scatter but can be represented to within 0.3% by the following equations:

$$a \text{ (in pm)} = 387.1 + 3.8 \times 10^{-3}\, t - 0.4 \times 10^{-6}\, t^2 \quad (t = 25 \text{ to } 1905°C) \text{ and}$$
$$c \text{ (in pm)} = 2738.5 + 28.4 \times 10^{-3}\, t + 6.2 \times 10^{-6}\, t^2 \quad (t = 25 \text{ to } 1905°C)$$

with the respective expansion coefficients α (in K⁻¹) = $9.82 \times 10^{-6} - 2.07 \times 10^{-9}\, t$ and γ (in K⁻¹) = $10.4 \times 10^{-6} + 4.53 \times 10^{-9}t$ (t = 25 to 1905°C) [8]. Pialoux [12] has made a single high-temperature determination of the lattice parameters of a carbon-saturated Th₃N₄ at about 1430°C obtaining values about 0.3% lower.

References for 4.1.4.2 see pp. 41/2

The c axis expands with increasing temperature more rapidly than the a axis. Correspondingly, the Th_2N_2 layer separation increases more rapidly with increasing temperature than the spreading of the layers. This expansion is very probably concentrated preferentially in the Th(I)–6N(II) (or Th(II)–1N(I)) bonds where the bond strength of the N atoms is least. An appreciable N deficiency in Th_3N_4 crystals at temperatures above 1950°C has been reported [13]. The formation of N vacancies which could account for this deficiency would seem to be energetically favored at these same sites.

Thermal Decomposition, Vapor Pressure

Pressures of N_2 in equilibrium with $Th_3N_4 + ThN$ mixtures at various temperatures have been reported by Aronson, Auskern [14], Benz et al. [13], Kusakabe, Imoto [15], and Benz [16]. The results representing the N_2 pressures at which Th_3N_4 decomposes by the reaction

$$2Th_3N_4 \rightleftharpoons 6ThN + N_2$$

are summarized by the equations

$$\log p \ (p \text{ in atm}) = -15888/T + 7.147 \ (T = 1723 \text{ to } 2073 \text{ K}) \ [14] \quad \dots\dots\dots \quad (9)$$

$$\log p \ (p \text{ in atm}) = -13990/T + 6.209 \ (T = 1723 \text{ to } 2173 \text{ K}) \ [15] \quad \dots\dots\dots \quad (10)$$

and

$$\log p \ (p \text{ in atm}) = -17800/T + 7.84 \ (T = 1673 \text{ to } 2073 \text{ K}) \ [16] \quad \dots\dots\dots \quad (11)$$

All three equations are in fair agreement at the more elevated temperatures of overlap but the p_{N_2} values calculated from the second equation are higher than those of the last equation at the lower temperatures of experimentation. The data represented by equation (10) were obtained by a dynamic technique with powder samples heated in a closed system as the temperature continuously increased while the decomposition was observed as a discontinuity in the optical emissivity at the Th_3N_4-to-ThN transformation. The data for equation (11) were obtained by a static technique of isothermal annealing. The cause of the differences between equations (10) and (11) below 1700 K is not clear. The decomposition reaction is sluggish at lower temperatures and any error from this is expected to give differences in the reverse direction. The Gibbs free energy change for decomposition of Th_3N_4 in 1 atm N_2 as calculated from equation (11) is

$$\Delta G \text{ (in cal/mol } N_2) = 81450 - 35.87T \ (T = 1673 \text{ to } 2073 \text{ K}).$$

The value $\Delta G = +17.85$ kcal/mol N_2 at 1773 K as calculated from this equation and the adopted C_p estimates of Rand [17] (see Table 10, p. 40) give with $S_{298}^\circ = 43.7$ cal·mol^{-1}·K^{-1} [20] by the 3rd-Law calculations for the enthalpy change of the reaction the value $\Delta H_{298} = 88.1$ kcal/mol. Combined with the enthalpy of formation of Th_3N_4, $\Delta H_{f,298}(Th_3N_4) = -309.6$ kcal/mol [18], the enthalpy of formation of ThN is $\Delta H_{f,298}(ThN) = -88.5$ kcal/mol, cf. p. 21. The temperature dependence of the Gibbs free energy change of the reaction $2Th_3N_4 \rightleftharpoons 6ThN + N_2$ can be represented by the equation

$$\Delta G \text{ (in cal/mol } N_2) = 87760 - 39.37T \ (T = 298 \text{ to } 2200 \text{ K}).$$

Enthalpy of Formation, Gibbs Free Energy of Formation

Reaction of Th filings made from 99% pure Th wire with N_2 in a bomb calorimeter gives for the enthalpy of formation of Th_3N_4 after correction to 1 atm N_2 $\Delta H_{f,298}(Th_3N_4) = -309.6 \pm 4.0$ kcal/mol $= -1295 \pm 17$ kJ/mol [18]. This value is based upon the assumption that all the N_2 uptake was present as Th_3N_4. The reaction was, in fact, only ⅔ complete. The presence of any

References for 4.1.4.2 see pp. 41/2

ThN, now known to exist and to form in appreciable amounts in incomplete reactions as a layer separating the Th metal and the outer Th_3N_4 product, was completely neglected. The enthalpy per gram atom N is appreciably greater for the formation of ThN than for Th_3N_4. For this reason, to allocate nitrogen to Th_3N_4 in the reaction product when it is, in fact, present as ThN, leads to a corresponding over-estimate of the absolute $\Delta H_f(Th_3N_4)$ magnitude. Apart from this uncertainty in the correction for ThN formation, this reaction calorimetric result is considered to be more direct and more reliable than the following ignition calorimetric results discussed and is deemed less susceptible to anomalies from oxygen contamination.

Measurements by standard O_2 ignition bomb calorimetry [19] give for the standard enthalpy of formation $\Delta H_{f,298}(Th_3N_4) = -312.4 \pm 4.4$ kcal/mol. This value, which was obtained as a difference of two measurements of larger quantities, agrees with the reaction calorimetric results (preceding paragraph) with the stated uncertainty, being only 2.8 kcal/mol greater. The coincidence is difficult to understand on the basis of the available information but an anomalously high ignition value could arise if the oxygen content of the nitride were underestimated.

In his review, Rand [17] chose an arithmetic average of the two different calorimetric values. Wagman et al. [21], in a recent review, give an even more negative value, -314.3 kcal/mol, presumably based on the same experimental data. For the reasons explained above, the more positive (lower absolute) of the two values is adopted here.

The $\Delta H_{f,298}$ values of all the other Th–N compounds, depending as they do on the reaction calorimetric value, are lower than the values tabulated by Rand [17]. To adopt a value different from the value $\Delta H_{f,298}(Th_3N_4) = -309.6$ kcal/mol, the $\Delta H_{f,298}$ values for all the other Th–N compounds have to be correspondingly adjusted by application of the first law of thermodynamics to the N_2 decomposition reaction equilibrium discussed here in the appropriate section for each of the compounds. The Gibbs free energy of formation of Th_3N_4 under standard conditions is represented to within 1.7 kcal/mol by the equation

$$\Delta G_f \text{ (in cal/mol } Th_3N_4) = -308\,000 + 80.58\,T \quad (T = 298 \text{ to } 2200 \text{ K)} \text{ [17]}.$$

This equation was calculated using the equation given above for the Th_3N_4 decomposition combined with 3rd Law calculations.

Table 10

Published Heat Capacity Data on Th_3N_4.

calorimetric method	temperature range in K	C_p (in cal·mol^{-1}·K^{-1}) = $a + b \times 10^{-3}T + c \times 10^6T^{-2}$			S°_{298} in cal·mol^{-1}·K^{-1}	Ref.
		a	b	c		
adiabatic	4 to 350				43.7	[20]
adiabatic	450 to 850	41.30	8.47	−0.637		[22]
estimated	298 to 2000	39.33	6.24	−0.533		[17]
estimated	298				42.7	[24]

Heat Capacity, Entropy

Reported heat capacity data are summarized in Table 10. The absolute entropy value $S^\circ_{298} = 43.7 \pm 0.5$ cal·mol^{-1}·K^{-1} as deduced by Dell, Martin [20] from their low-temperature heat content measurements is adopted here. Wagman et al. [21] state the value $S^\circ_{298} = 48$ cal $\cdot$ mol$^{-1}\cdot$K^{-1}, presumably estimated. The high-temperature heat capacity of Th_3N_4 as estimated by Rand [17] is given by the equation

$$C_p \text{ (in cal}\cdot\text{mol}^{-1}\cdot\text{K}^{-1}) = 39.33 + 6.24\times10^{-3}T - 0.533\times10^6T^{-2}.$$

The values measured by Ono et al. [22] are higher by up to 9%. The "Th_3N_4" on which the ice calorimetric measurements between 273 and 773 K were made by Satoh [23] is today no longer considered satisfactorily characterized – the material contained 2.65% oxygen, the chemical form of which is unknown.

Electrical Resistivity

This compound is an insulator [27]. The room-temperature specific dc electrical resistivity of hot-pressed Th_3N_4 with 18% porosity after annealing in 1 atm N_2 at 1850°C for 6 h is reported as being 10^3 to 10^6 $\Omega\cdot$m and having a negative temperature coefficient [28]. Later Kamegashira et al. [29] reported results of 4-probe measurements made on 56% dense Th_3N_4 after sintering for 10 h in 1 atm N_2 at 1350°C. They are presented as a plot of log σ vs. 1/T, where σ (in $\Omega^{-1}\cdot$m^{-1}) is the specific conductivity. The plot shows a weak curvature with the mean activation energy $E_A = 1.40$ eV $= 32400$ cal/mol and can be represented by the equation

$$\log \sigma \text{ (in } \Omega^{-1}\cdot\text{m}^{-1}) = -7057/T - 0.9116\log T + 8.143 \quad (T = 1000 \text{ to } 1353 \text{ K}).$$

The dependence of the conductivity on p_{N_2} at the temperatures 1273, 1323, 1373, 1423, and 1473 K was measured by Kamegashira et al. [29] and found to increase with decreasing N_2 pressure in the range 10^2 to 10^5 Pa $= 10^{-3}$ to 1 atm, suggesting an n-type semiconductor. Slopes of log σ vs. log p_{N_2} plots were reported to be $-1/8.37$, $-1/9.80$, $-1/8.73$, $-1/7.96$, and $-1/8.50$, respectively. Positive deviations from straight line plots occur at pressures approaching 1 atm N_2, however, and the deviations are greater at lower temperatures. The predominant defects are suggested to be "ionized N vacancies".

Superconductivity

No superconducting transition occurs in pure Th_3N_4 at 6 K, in agreement with the general rule that ionic compounds are not superconductors [30].

Magnetic Susceptibility

A temperature-independent paramagnetism has been reported but the χ values are extremely small, ranging from $+3$ to 6×10^{-9} (83 to 295 K), [25] and $+50\times10^{-9}$ cm^3/g [26], Adachi, Imoto [26] observed the presence of a small ferromagnetic component at low fields that could not be explained, but this may be due to the presence of impurity.

Color

Th_3N_4 is maroon but changes on sintering to a very dark brown to gray [7, 14].

References for 4.1.4.2:

[1] Benz, R., Zachariasen, W. H. (Acta Cryst. 21 [1966] 838/40).
[2] Bowman, A. L., Arnold, G. P. (Acta Cryst. B 27 [1971] 243/4).

[3] Intern. Tables for X-Ray Crystallogr., Vol. 1, Kynoch Press, Birmingham, Engl., 1965.
[4] Benz, R., Arnold, G. P., Zachariasen, W. H. (Acta Cryst. B **28** [1972] 1724/7).
[5] Benz, R. (J. Am. Chem. Soc. **89** [1967] 197/9).
[6] Juza, R., Gerke, H. (Z. Anorg. Allgem. Chem. **363** [1968] 245/57).
[7] Chioti, P. (AECD-3072 [1951] 1/63; N.S.A. **5** [1951] 3141).
[8] Benz, R., Balog, G. (High Temp. Sci. **3** [1971] 511/22).
[9] Kusakabe, T., Imoto, S. (Nippon Kinzoku Gakkaishi **35** [1971] 795/800).
[10] Kusakabe, T., Imoto, S. (Technol. Rept. Osaka Univ. **22** [1972] 477/88).

[11] Zachariasen, W. H. (in: Seaborg, G. T., Katz, J. J., The Actinide Elements, Crystal Chemistry of the 5f Elements, McGraw-Hill, New York 1954, pp. 769/96).
[12] Pialoux, A. (J. Nucl. Mater. **91** [1980] 127/46).
[13] Benz, R., Hoffman, C. G., Rupert, G. N. (J. Am. Chem. Soc. **89** [1967] 191/7).
[14] Aronson, S., Auskern, A. B. (J. Phys. Chem. **70** [1966] 3937/41).
[15] Kusakabe, T., Imoto, S. (Nippon Kinzoku Gakkaishi **35** [1971] 1115/20).
[16] Benz, R. (J. Electrochem. Soc. **119** [1972] 1596/602).
[17] Rand, M. H. (At. Energy Rev. Spec. Issue No. 5 [1975] 7/85).
[18] Neumann, B., Kröger, C., Haebler, H. (Z. Anorg. Allgem. Chem. **207** [1932] 145/9).
[19] Neumann, B., Kröger, C., Kunz, H. (Z. Anorg. Allgem. Chem. **218** [1934] 379/401).
[20] Dell, R. M., Martin, J. F. (At. Energy Rev. Spec. Issue No. 5 [1975] 7/85).

[21] Wagman, D. D., Evans, W. H., Parker, V. B., Schumm, R. H. Nuttall, R. L. (NBS-TN-270-8 [1981] 1/53; C.A. **69** [1982] No. 12232).
[22] Ono, F., Kanno, M., Mukaibo, T. (J. Nucl. Sci. Technol. [Tokyo] **10** [1973] 391/5).
[23] Satoh, S. (Sci. Papers Inst. Phys. Chem. Res. [Tokyo] **35** [1939] 182/90).
[24] Kelley, K. K. (U.S. Bur. Mines Bull. No. 407 [1937] 1/66; C.A. **1938** 2821).
[25] Aronson, S., Auskern, A. B. (J. Chem. Phys. **48** [1968] 1760/5).
[26] Adachi, H., Imoto, S. (Technol. Rept. Osaka Univ. **23** [1973] 1121/54, (Engl.) 425/9).
[27] Düsing, W., Hüniger, M. (Tech. Wiss. Abhandl. Osram-Ges. **2** [1931] 357/65).
[28] Auskern, A. B., Aronson, S. (J. Phys. Chem. Solids **28** [1967] 1069/71).
[29] Kamegashira, N., Tsuji, T., Miyamoto, T., Naito, K. (J. Nucl. Mater. **102** [1981] 26/9).
[30] Giorgi, A. L., Szklarz, E. G. (Private communication 1970).

4.1.4.3 Chemical Behavior

For thermal decomposition, see p. 39.

The reaction $Th_3N_4 + 3O_2 \rightarrow 3ThO_2 + 2N_2$ is very slow below 340°C, being incomplete even after 2 d. The time course of reaction is represented empirically by an equation of the same form as for ThN, see p. 34. The energy of activation, 15.2 ±1.7 kcal/mol (340 to 480°C), does not differ significantly from that found with ThN. Some self-heating but no ignition of Th_3N_4 occurs in 0.2 atm O_2 at 480°C. The rate-controlling oxidation mechanism in Th_3N_4 is the same as for ThN [1]. The reactivity with moist and with dry air is similar to that described for ThN. Reaction of Th_3N_4 at 0 to 50°C with air of 20% humidity yields an amorphous ThO_2, which can be converted to various degrees of crystallinity by heat treatment of the product ThO_2 at 200 to 1000°C [2].

The reaction with liquid water, $Th_3N_4 + 6H_2O \rightarrow 3ThO_2 + 4NH_3$, proceeds at roughly the same rate as that of ThN; however, the rate of reaction does not follow the equation $\log (c/(c - c_o)) = kt$ [3].

The brown powder of metastable β-Th_3N_4 is stable in 1 atm N_2 to 1020°C, hydrolyzes in most air, and readily dissolves in dilute HCl with no gas evolution [4].

References for 4.1.4.3:

[1] Ozaki, S., Kanno, M., Mukaibo, T. (J. Nucl. Sci. Technol. [Tokyo] **8** [1971] 414).
[2] Miyake, M., Katsura, M. (Proc. Rept. Spec. Proj. Res. Energy, Osaka 1984, SPEY 9, pp. 105/8).
[3] Sugihara, S., Imoto, S. (J. Nucl. Sci. Technol. [Tokyo] **8** [1971] 630/6).
[4] Juza, R., Gerke, H. (Z. Anorg. Allgem. Chem. **363** [1968] 245/57).

4.2 Ternary Nitrides of Thorium with Other Metals

Benz, R., Naoumides, A.
Kernforschungsanlage Jülich
Jülich, Federal Republic of Germany

4.2.1 Overview and General Remarks

The subsequently discussed ternary nitrides are mostly high-temperature materials. Preparation reactions of multicomponent Th–N–X mixtures must allow for very slow chemical interdiffusion coefficients in solids. Many preparations can be made by powder metallurgical techniques involving multiple comminution, mixing, and elevated-temperature annealing, e.g., formation of (Th,U) N solid solutions from the binary compounds.

Chemical analytical techniques for Th, N, C, and O as previously described for the binary compounds (see p. 9) are applicable to polynary alloys except that attention must be given to cations that can interfere with analyses for Th [1 to 5]. With colorimetric Th analyses, the selectivity is generally greater for Th^{4+} in strong acids. Thus, the coloring agent Thoron with its optimum pH range 0.3 to 1 can react more specifically with Th^{4+} than Morin in the pH range 1.8 to 2 within certain limits even in the presence of the interfering ions of U, Zr, Ti, lanthanides, and Fe [3].

Dissolution of ternary and higher alloys is sometimes difficult. Analysis for N, in such cases, may be expedient by the Dumas method whereby the sample is fused in an oxidizing fluxing agent and the off-gases absorbed in NaOH solution and the residual gas analyzed gas volumetrically as N_2.

References for 4.2.1:

[1] Banks, C. V., Diehl, H. (Ind. Eng. Chem. Anal. Chem. **19** [1947] 222/4).
[2] Körbl, J., Přibil, R., Emr, A. (Collection Czech. Chem. Commun. **22** [1957] 961/6).
[3] Lange, B. Vejdelek, Z. J. (Photometrische Analyse, Verlag Chemie, Weinheim 1980, pp. 325/30).
[4] Rodden, C. J., Warf, J. C. (Natl. Nucl. Energy Ser. Div. VIII **1** [1950] 160/207).
[5] Ryabchikov, D. I., Gol'braikh, E. K. (in: Analytical Chemistry of Thorium, Ann. Arbor Humphrey, Ann. Arbor, Mich., 1969, pp. 1/289, 16/73).

4.2.2 Ternary Nitrides of Thorium with Main Group Metals (Li, K, Be, Mg, Sb, Bi)

4.2.2.1 The Th–Li–N System

Li_2ThN_2. The compound is prepared by slowly heating a mixture of Li_3N and Th or Li_3N and Th_3N_4 with ca. 30% excess Li_3N to compensate for vaporization losses [1]. The powder mixture is slowly heated in a stream of NH_3 to above 300°C where the intermediate compound $LiNH_2$ forms, to 380°C where some melting takes place, and then to 500 to 600°C where a strong evolution of H_2 occurs, and, finally for 1 to 2 h at 700 to 800°C [1]. Barker, Alexander [2] formed the compound by heating any of the mixtures $1.7\,Li_3N + Th$ in N_2 for 1 d at 600°C and at 640°C, $3\,Li_3N + Th$ in vacuum for 1 d at 590°C, and $2\,Li_3N + ThN$ in NH_3 for 1 d at 570°C.

Crystals of Li_2ThN_2 are hexagonal with a = 639.8 and c = 554.7 pm and the theoretical density $D_X = 6.950$ g/cm³ [1] and a = 638.8 and c = 553.6 pm corresponding to $D_X = 6.98$ g/cm³ [2]. The structure according to Palisaar, Juza [1]: Space group $P\bar{3}$-C_{3i}^1 (No. 147) with 3 formula units per unit cell.

Atomic positions: $1\,Th(I)$ in $(0,0,0)$; $2\,Th(II)$ in $\pm\,(^1/_3,\,^2/_3,\,z)$; $6\,Li$ and $6\,N$ in $\pm\,(x, y, z)\,(\bar{y}, x - y, z)$ $(y - x, \bar{x}, z)$ with the position parameters:

	x	y	z
Th^{4+}			0.020
N^{3-}	0.34	0.01	0.27
Li^+	0.35	0.03	0.63

The structure has alternating hexagonal layers of 3 Th and $3\,(N_2Li_2)$ per unit cell, the latter layer forming a 2-dimensional array of Li–4N tetrahedra. The N^{3-} are in a nearly close-packed hexagonal configuration with the Th^{4+} ions occupying octahedral and the Li^+ ions occupying tetrahedral holes [1]. The following interatomic distances (in pm) have been calculated by the authors of this article on the basis of the given description in [1]:

Th(I)–6 Th(II)	370	Th(I)–6 N	362	Th–6 Li	297
Th(II)–3 Th(I)	370	Th(II)–3 N	258	Th–3 Li	278
		Th(II)–3 N	263		

N–1 Th(I)	262	N–1 Li	200
N–1 Th(II)	258	N–1 Li	212
N–1 Th(II)	263	N–1 Li	219
		N–1 Li	231

The yellow-to-gray compound Li_2ThN_2 hydrolyzed in moist air with NH_3 evolution [1]. The pycnometric density is reported as being D = 6.84 [1] and 6.76 g/cm³ [2].

4.2.2.2 The Th–K–N System

$K_3Th_3N_5$. The relatively stable composition $K_3Th_3N_5$ is formed by thermal decomposition of any of the three compounds $K_2Th_2(NH_2)_{2n}(NH)_{5-n}$ (n = 1, 2, or 3) by slowly heating in a stream of N_2 at temperatures above 150°C [3]. The brown-yellow amorphous compound hydrolyzes in water, decomposes to ThN in a stream of N_2 at 425°C, and explodes violently in air [3].

References for 4.2.2 see p. 48

4.2.2.3 The Th–Be–N System

BeThN₂. The compound $BeThN_2$ has been prepared by Palisaar, Juza [1] by heating an equimolar mixture of Be_3N_2 and Th_3N_4 pressed into the form of a tablet and supported on Ta in 1 atm N_2. It is heated for 7 to 45 min at 1400 to 1880°C. The gray-to-olive colored product is hard and well crystallized. It has a pycnometric density D (25°C) $= 9.1^{+0.10}_{-0.03}$ g/cm³. It decomposes in 1 atm N_2 above 1880°C. It is stable in water and in air but soluble in dilute HCl.

Crystals of $BeThN_2$ have been reported to have the hexagonal unit cell with a = 1050.1 and c = 395.5 pm, theoretical density D_X = 9.464 g/cm³, and 8 formula units per cell [1].

4.2.2.4 The Th–Mg–N System

An attempt to prepare a Mg compound analogous to the $BeThN_2$ was unsuccessful [1].

4.2.2.5 The Th–Sb–N System

Phase Relations

Qualitative features of the Th–Sb–N 1273-K isotherm are illustrated in **Fig. 6**. This phase diagram differs from that of the analogous Th–N–As and Th–N–P systems in containing the tetragonal compound Th_2N_2X with X = Sb. This compound forms a continuous series of solid solutions with the isostructural compound Th_2NOSb, whereas the hexagonal structures Th_2NOP and Th_2NOAs are essentially line compounds. It differs from the Th–Bi–N 1273-K isotherm which does contain the ternary compound with X = Bi in that no fcc ThBi compound exists [4] so that the solubility of Bi in ThN is limited.

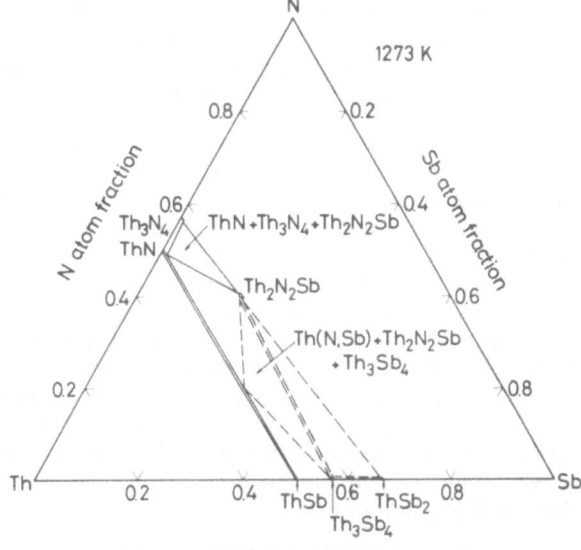

Fig. 6

Isothermal section in the Th–Sb–N system at 1273 K, suggested from cited literature.

References for 4.2.2 see p. 48

Th₂N₂Sb

Preparation. Th_2N_2Sb has been synthesized in two different ways: 1) the cold-pressed powder mixture ThSb + ThN is heated in a tungsten crucible and lid assembly under 1 atm N_2 for 2 h at 1200°C and 2) the powder mixture 2ThN + Sb is sealed in an evacuated quartz capsule and heated for 30 d at 1000°C [5].

Crystal Structure. Body-centered tetragonal, U_2N_2Sb-type, Th_2N_2Sb crystals have the lattice parameters a = 404.9 ± 0.1 and c = 1357 ± 1 pm with the theoretical density D_X = 9.17 g/cm³ [5]. The crystal structure is: Space group I4/mmm-D_{4h}^{17} (No. 139); atomic positions (0,0,0)(½, ½, ½) + 2Sb in (0,0,0), 4Th in ±(0,0,z_{Th}), z_{Th} = 0.344 ± 0.003; 4N in ±(½, 0, ¼). The interatomic distances (in pm, calculated by the authors of this article, based on the data in [5]) are: Th–Th 383 ± 3 (4×), Th–N 239 ± 3 (4×), Th–Sb 356 ± 3 (4×), and N–Th 239 (4×). The configuration of the 4Th atoms about an N atom is tetragonal. Four of the edges of each NTh₄ tetrahedron are shared with other tetrahedra so as to produce endless layers normal to the c axis of composition Th_2N_2 per unit area a². The shared edges of the tetrahedron are appreciably shorter (368.5 pm) than the unshared edges (404.9 pm). Each Sb atom is bonded to 8Th atoms which form a nearly perfect cube, the edges being 404.9 pm normal to the c axis and 423 pm parallel to it. The structure can be described as a stack of tetrahedral layers Th_2N_2 with alternating Sb layers in the sequence $[Th_2N_2]Sb[Th_2N_2]*Sb*[Th_2N_2]Sb$ ... where the asterisk denotes the horizontal displacement $(½)(a_1 + a_2)$ [5].

The structure is representative of the isostructural compounds represented by the general formula $Th_2(N,O)_2X$, X = Sb, Bi, and Te. The lattice parameters of Th_2N_2Te (cf. "Thorium" C 5, 1986, p. 115) show no significant variation in the presence of oxygen. However, with increasing O contents the length of the c axis in the other two compounds decreases. The observed range of values is 1357 to 1284 ± 1 pm in $Th_2(N,O)_2Sb$ and 1362 to 1353 ± 3 pm in $Th_2(N,O)_2Bi$ preparations. This is remarkable because bond length formed by an O atom is usually only slightly smaller than that of an N atom. These results can be understood in terms of Zachariasen's bond strength, recalling that in the O-lean compounds with X = Sb and Bi, divalent X^{2-} anions must exist to satisfy the Th^{4+} and N^{3-} valence requirements. At the composition $Th_2(N_{1/2}O_{1/2})_2Sb$ the strength of each Th–Sb bond is 0.375; but, in the compound Th_2N_2Sb the strength of the Th–Sb bond has decreased to only 0.25 [5].

The observed U–N = 227 ± 2 pm in U_2N_2Sb [5] is much smaller than expected based on Th–N = 239 pm in Th_2N_2Sb, while both the M–Sb (M = U, Th) are larger than anticipated in the M_2N_2Sb compounds. For example the distance Th–Sb is 356 pm in Th_2N_2Sb as compared to 316 pm in ThSb [6] and U–Sb is 336 pm in U_2N_2Sb as compared to 310 pm in USb [7]. The apparently anomalous distances in the M_2N_2Sb compounds can be attributed to unequal distribution of strengths between the M–N and M–Sb bonds. Since each N atom forms 4 bonds to the M atoms, the strength of each bond must be 0.75 in order to satisfy the valence of 3 for N. Thus, one should expect the same M–N distance as in the respective compounds Th_3N_4 and UN_2 where the bond strengths are 0.75. The bond distances are Th–N = 234 in Th_3N_4 and U–N = 230 pm in UN_{2-x} [8]. Similarly, the bond strength of each M–Sb bond is 0.375 in the U_2N_2Sb and 0.25 in Th_2N_2Sb and an appreciably larger M–Sb bond length is expected than in MSb where the bond strength is 0.5. The large difference of 20 pm between the Th–Sb and the U–Sb bond lengths is mainly due to the difference in the Sb effective valence. The valence is Sb^{3-} in U_2N_2Sb and Sb^{2-} in Th_2N_2Sb [5].

The ternary compounds with the general formula $Th_2(N,O)_2X$, where X denotes the heavy elements below row 1 in Columns V and VI of the periodic table, crystallize in two different structure types, a hexagonal and a body-centered tetragonal one. The crystal type assumed depends on the size of the X atom [5]. When X = Sb, Bi, or Te, the X atom can have a

References for 4.2.2 see p. 48

coordination number of 8 with respect to Th without bringing the Th atoms in too close contact and the tetragonal structure results. However, with the smaller atoms X = P, As, S, or Se, the coordination number for the atom X must be reduced from 8 to 6 in order to avoid too short Th–Th separations, and the Th_2N_2S-type hexagonal structure prevails [5] (cf. the compounds Th_2N_2S and Th_2N_2Se, "Thorium" C 5, 1986, p. 53 and p. 104).

4.2.2.6 The Th–Bi–N System

Phase Relations

Qualitative features of the 1273-K isotherm as illustrated in **Fig. 7** are suggested based upon the available information in [5] on the binary and ternary mixtures. The ThBi compound evidently does not form an fcc structure [4] therefore, the solubility in fcc ThN must be limited.

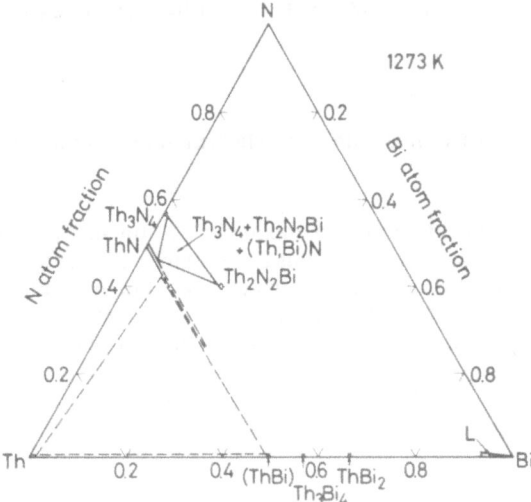

Fig. 7
Isothermal section in the Th–Bi–N system at 1273 K, suggested on the basis of cited literature.

Th_2N_2Bi

Preparation. The compound Th_2N_2Bi in which an appreciable amount of the N (up to ca. 13%) can be replaced by dissolved oxygen is prepared in either of two ways: 1) cold-pressed powder mixtures of the binary compounds ThN + ThBi or of $x ThO_2 + (1-x)ThN + ThBi$ with x ranging from 0 to about $^1/_8$ are heated in a W crucible with lid under 1 atm N_2 for 16 h at 1000°C or 2) the mixture 2 ThN + Bi or $x(ThO_2 + Th) + 2(1-x)ThN + Bi$ is equilibrated in a silica capsule for 30 d at 1000°C [5].

Crystal Structure. Th_2N_2Bi forms body-centered tetragonal, U_2N_2Sb-type, crystals isostructural with Th_2N_2Sb. The lattice parameters decrease upon exchanging O for N atoms from $a = 407.5 \pm 0.1$ and $c = 1362 \pm 1$ in Th_2N_2Bi to $a = 407.4 \pm 0.2$ and $c = 1353 \pm 3$ pm in oxygen-

References for 4.2.2 see p. 48

saturated $Th_2(N,O)_2Bi$. The respective theoretical densities are $D_x = 10.30$ and 10.37 g/cm³ [5]. The interatomic distances of Th_2N_2Bi (in pm) are: Th–Th 385 ± 10 (4×), Th–N 241 ± 8 (4×), Th–Bi 358 ± 8 (4×), and N–Th 241 (4×) [5].

References for 4.2.2:

[1] Palisaar, A.-P., Juza, R. (Z. Anorg. Allgem. Chem. **384** [1971] 1/11).
[2] Barker, M. G., Alexander, I. C. (J. Chem. Soc. Dalton Trans. **1974** 2166/70).
[3] Schmitz-Dumont, O., Raabe, F. (Z. Anorg. Allgem. Chem. **277** [1954] 297/314).
[4] Ferro, R. (Acta Cryst. **10** [1957] 476/7).
[5] Benz, R., Zachariasen, W. H. (Acta Cryst. B **26** [1970] 823/7).
[6] Ferro, R. (Acta Cryst. **9** [1956] 817/8).
[7] Ferro, R. (Atti Accad. Nazl. Lincei, Rend. Classe Sci. Fis. Mat. Nat. Rend. [8] **13** [1952] 53/61).
[8] Pearson, W. B. (Lattice Spacings and Structures of Metals and Alloys, Vol. 2, Pergamon, New York 1967, pp. 1/1456; C.A. **67** [1967] No. 14412).
[9] Darnell, A. J., McCollum, W. A., Milne, T. A. (J. Phys. Chem. **64** [1960] 341/6).

4.2.3   Ternary Nitrides of Thorium with Group III Transition Metals (Rare Earths, Actinides)

The Th–Y–N System

Alloys prepared by arc melting of Th-Y mixtures in 3 atm, and subsequently homogenization in N_2 at 1900 to 2000 K show a continuous series of fcc solid solutions existing between YN and ThN. The room-temperature lattice parameters vary linearly with metal atom ratio [1]. A miscibility gap with the critical point of 750 K has been estimated by Holleck, Shohoji [2]. The 2073-K isotherm as postulated by Holleck [3] is illustrated in **Fig. 8**.

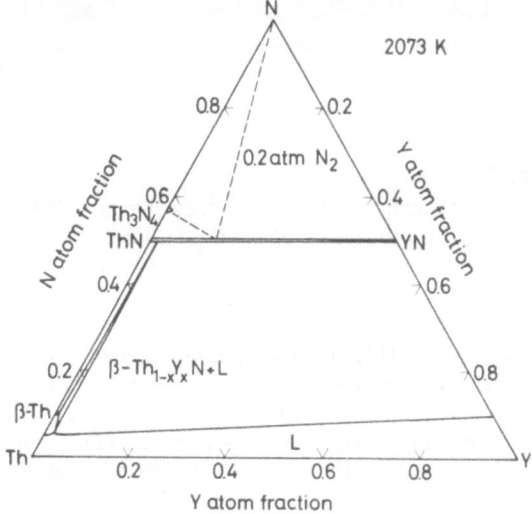

Fig. 8

Isothermal section in the Th–Y–N system at 2073 K, suggested on the basis of the cited literature.

References for 4.2.3 see p. 50

The Th–La–N System

A continuous series of solid solutions forms between fcc LaN and ThN at 1700°C [4] and at 1900°C [1] with only a small negative deviation from the Vegard rule in LaN-rich solutions. No ternary compound exists at 1200°C and N_2 pressures up to 30 atm, at which pressure the solid solution decomposes into the binary nitrides Th_2N_3 (sic, probably Th_3N_4) and LaN [4]. A miscibility gap is predicted with the calculated critical point of 1450 K [2].

The Th–Ce–N System

A continuous series of fcc solid solutions form between CeN and ThN with positive deviations from the Vegard rule [1, 4]. The 1973-K isotherm proposed by Holleck [3] is illustrated in **Fig. 9**.

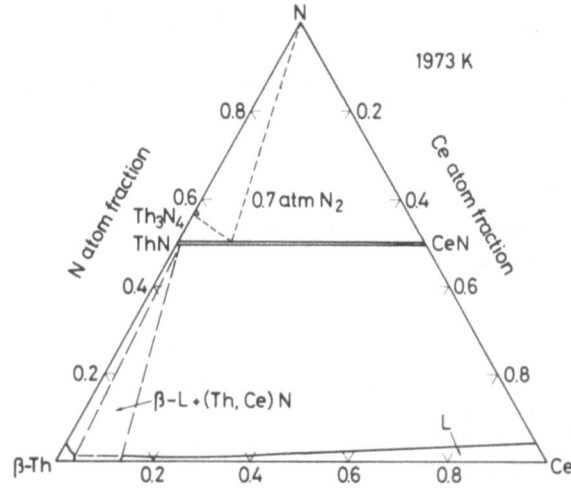

Fig. 9

Isothermal section in the Th–Ce–N system at 1973 K, suggested on the basis of the cited literature.

The Th–Pr(Nd,Sm,Gd,Dy)–N Ternary Systems

A continuous series of fcc solid solutions form between MN (M = Pr, Nd, Sm, Gd, Dy) and ThN with negative deviations from the Vegard rule (in the case of Pr and Nd only small) after quenching from 1700°C [4]. According to Holleck, Smailos [1] the room temperature lattice parameters of Th–Pr–N and Th–Nd–N solid solutions are reported to follow the Vegard rule after quenching from 1900°C.

The Th–U–N System

ThN and UN are completely miscible and form a continuous series of solid solutions. Phase relations, preparation and properties (crystallographic properties, thermal conductivity, electrical and magnetic properties) of the solid solutions are described in the Gmelin Handbook, see "Uranium" Suppl. Vol. C 7, 1981, pp. 81/2.

References for 4.2.3 see p. 50

Some supplementary statements concerning nuclear fuel performance properties: Breeding ratios of homogeneous and of heterogeneous (Th, U) N cores (fully enriched ^{15}N) with fuel pins having a 9.40-mm diameter and clad with 0.38-mm thick 316 stainless steel in 1200 MW (elec.) LMFBR's (liquid-metal fast breeder reactor) were calculated for the peak linear power of 98.4 kW/m to be 1.16 and 1.24, respectively [5]. The compound inventory doubling times were calculated to be 53 and 46y, respectively, as compared to 82 and 53y, respectively, for corresponding carbide fuel. A 510°C mixed mean reactor outlet temperature gives 36% thermal efficiency [5]. Calculations for heterogeneous cores with similar design assumptions have confirmed that Th–U nitride fuel performance is better than that of carbide fuel and equals that of metal fuels [6]. The lowest doubling time ranges from 30 to 36y with specific inventories of 3.6 to 4.3 kg fissile/MW(e).

The Th–Pu–N System

A complete series of solid solutions forms between PuN and ThN [1, 3, 7]. An alloy with the composition $(Th_{0.15}Pu_{0.85})N$ and containing 0.1 to 0.3% oxygen has been reported to be single phase with the fcc lattice parameter a = 498.0 pm indicating an appreciable positive deviation from the Vegard rule [8].

The Th–Np–N and Th–Am–N Systems

A continuous series of fcc solid solutions between ThN and NpN [1, 3] and AmN, respectively [1] is reported to be probable.

References for 4.2.3:

[1] Holleck, H., Smailos, E. (J. Nucl. Mater. **91** [1980] 237/9).
[2] Holleck, H., Shohoji, N. (cited in [1]).
[3] Holleck, H. (Thermodyn. Nucl. Mater. Proc. 4th Symp., Vienna 1974 [1975], Vol. 2, pp. 213/63; C.A. **84** [1976] No. 5040).
[4] Ettmayer, P., Waldhart, J., Vendl, A., Banik, G. (Monatsh. Chem. **111** [1980] 945/8).
[5] Caspersson, S. A., Kulwich, M. R. (Advan. React. Phys. Proc. Am. Nucl. Soc. Topical Meeting, Gatlinburg, Tenn., 1978, pp. 539/48).
[6] Turski, R. B., Lam, P. S. K., Barthold, W. P. (Trans. Am. Nucl. Soc. **28** [1978] 563/5).
[7] Holleck, H. (KFK-1726 [1972] 1/29; C.A. **79** [1973] No. 108276).
[8] Pardue, W. M., Storhok, V. W., Smith, R. A. (Proc. 3rd Intern. Conf. Plutonium, London 1965 [1967], pp. 721/38).

4.2.4   Ternary Nitrides of Thorium with Group IV and Group V Transition Metals
(Ti, Zr, Hf, V, Nb, Ta)

Th–Ti–N. A ThN + 25 wt% Ti alloy, after sintering for 2 h at 500°C and then at 1630°C in 6×10^{-6} atm N_2 for 1 h, is reported to have a Vickers hardness of 1240 and a transverse rupture strength of 112×10^6 kg/m^2 ($= 11 \times 10^8$ N/m^2) [1]. Solubility of fcc TiN in ThN is predicted to be limited [2, 3].

Th–Zr–N. Some qualitative features of the 1773-K isotherm as calculated by Holleck [3] and based upon experimental work attributed by him to Chubb, Keller are illustrated in **Fig. 10.** A continuous series of solid solutions is shown to exist between ZrN and ThN. The solution separates on the metal-rich side of the ThN–ZrN isopleth into a miscibility gap between ThN and fcc ZrN_{1-x}. The solubility between the fcc mononitrides ZrN and ThN has been predicted to be limited (presumably at low temperatures) [2, 3].

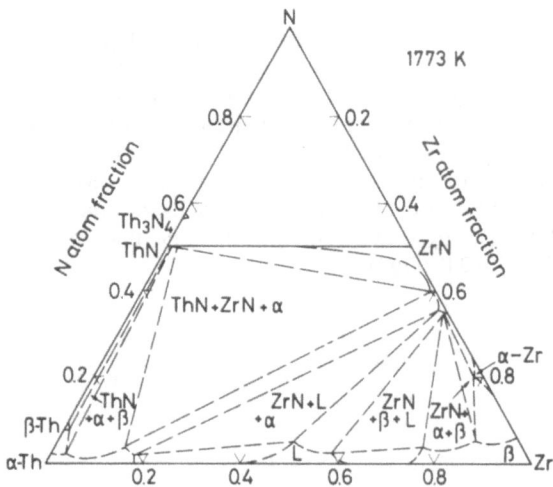

Fig. 10

Isothermal section in the Th–Zr–N system at 1773 K, suggested on the basis of cited literature.

Th–Hf–N. A limited solid solubility between fcc HfN and ThN is predicted [2, 3].

Th–V–N. No ternary compound in this system has been reported although an orthorhombic UVN compound in the analogous U–V–N system has been claimed [4, 5]. A limited solubility of fcc VN in ThN is predicted [2, 3].

Th–Nb–N. An incomplete solubility of fcc NbN in ThN is predicted [2].

Th–Ta–N. Solubility between fcc TaN and ThN is predicted to be limited [3].

References for 4.2.4:

[1] Tomita, C., Kasai, I., Kohiki, T. (Japan. 76-46721 [1976]; C.A. **87** [1977] No. 57168).
[2] Gusev, A. I., Shveikin, G. P. (Izv. Akad. Nauk SSSR Neorgan. Materialy **10** [1974] 2144/7; Inorg. Materials [USSR] **10** [1974] 1840/3), see also Gusev, A. I. (Izv. Akad. Nauk SSSR Neorgan. Materialy **19** [1983] 1319/24; Inorg. Materials [USSR] **19** [1983] 1183/8).
[3] Holleck, H. (Thermodyn. Nucl. Mater. Proc. 4th Symp., Vienna 1974 [1975], Vol. 2, pp. 213/63; C.A. **84** [1976] No. 5040).
[4] Holleck, H. (KFK-1011 [1969] 1/13; N.S.A. **24** [1970] No. 3023).
[5] Holleck, H. (KFK-1726 [1972] 1/29; C.A. **79** [1973] No. 108276).

4.2.5 Ternary Nitrides of Thorium with Groups VI, VII, and Group VIII Transition Metals (Cr, W, Mn)

4.2.5.1 The Th–Cr–N System

Phase Relations

General features of the 1873-K isotherm as illustrated in **Fig. 11** show the existence of the ternary compound Th_2CrN_3 coexisting with Cr(+N) metal which, with the assumption of negligible solubility of Th, agrees with the extrapolated N_2 decomposition pressure data on Cr_2N [1]. Holleck [2] has proposed a phase diagram according to which the products of decomposition are $ThN + Cr_2N + N_2$ at 1473 K instead of $ThN + Cr(+N) + N_2$ as illustrated in Fig. 11. The subsequently discussed thermodynamic data give $p_{N_2} = 4 \times 10^{-5}$ atm for the Th_2CrN_3 decomposition pressure at 1473 K. This is lower than the Cr_2N decomposition pressure given by Mills [1], $p_{N_2} = 0.06$ atm N_2, indicating that the 1473-K isotherm also is qualitatively like that at 1873 K (Fig. 11).

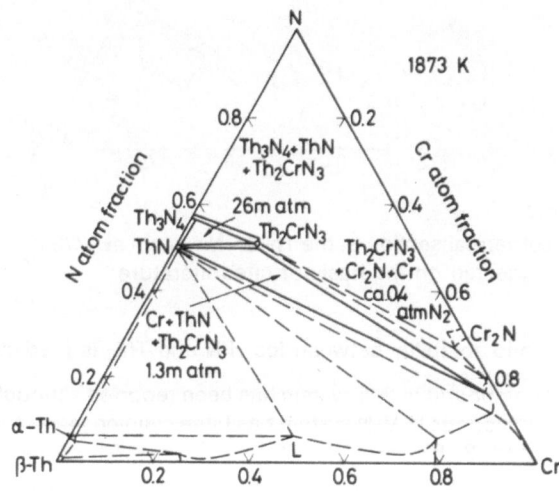

Fig. 11

Isothermal section in the Th–Cr–N system at 1873 K, suggested by the authors of this article on the basis of published data on binary phase diagrams [14] and discussed decomposition pressures.

Th_2CrN_3

Preparation. The compound Th_2CrN_3 is prepared by equilibrating cold-pressed powders $2ThN + Cr$ in 1 atm N_2 at 1600°C in a W crucible [3]. Another technique that should be applicable and that has been used for the preparation of the isostructural compound U_2CrN_3 starts with arc melting of $2Th + Cr$ in 3 atm N_2 followed by homogenization at 1300°C in N_2 [4, 5]. However, the latter technique has the disadvantage that it is difficult to predict Cr vaporization losses and the vaporization loss can be expected to be greater for the $2Th + Cr$ than for the $2U + Cr$ alloy.

References for 4.2.5 see p. 56

Crystal Structure. Orthorhombic U_2CrN_3-type crystals of Th_2CrN_3 have the lattice parameters a = 386.54 ± 0.05, b = 351.54 ± 0.02, c = 1284.46 ± 0.19 pm and the theoretical density D_x = 10.623 ± 0.004 g/cm³ [3]. The structure having 2 formula units per unit cell is: Space group Immm-D_{2h}^{25} (No. 71); atomic positions (0,0,0) (½, ½, ½) + 2Cr, 4Th in ±(0, 0, z_{Th}), z_{Th} = 0.356 ± 0.003, 2N(I) in (½, 0, 0), 4N(II) in ±(0, 0, z_N), z_N = 0.151 ± 0.003. The structure is also formed by Th_2MnN_3. The interatomic distances (in pm) are: Th–Th 351.54 (2×), Th–N(I) 255 ± 3 (2×), Th–N(II) 261.4 ± 0.5 (4×), Th–N(II) 263 ± 7 (1×); Th–Cr 320 ± 2 (4×); N(I)–Th 255 (4×), N(II)–Th 261.4 (4×), N(II)–Th 263 (1×); Cr–N(I) 193.3 (2×); Cr–N(II) 194 ± 4 (2×); N(I)–Cr 193.3 (2×), and N(II)–Cr 194 (1×). Each Cr atom forms 4 Cr–N bonds directed toward the corners of a square. Two opposing corners of each square are shared with adjacent squares to form endless CrN_3 chains along the $\bar{a}$ axis. These endless CrN_3 chains are linked together by means of Cr–N bonds, such that each N(I) atom is bonded to two Cr atoms and to 4 Th atoms while each N(II) atom is bonded to one Cr atom and to 5 Th atoms. Along the $\bar{b}$ axis there are rather strong Th–Th bonds [3].

The observed Th–N and Cr–N bond lengths are nearly the same as observed in the binary nitrides ThN and CrN calculated from published data as 258 pm in ThN and 207 pm in CrN [6].

Thermodynamic Properties. The N_2 decomposition pressure of Th_2CrN_3 in the presence of excess Cr has been measured as being 1 Torr at 1873 K for the reaction $2\,Th_2CrN_3 \rightleftharpoons 4\,ThN + 2\,Cr + N_2$. The corresponding free energy of reaction is $\Delta G = -RT \cdot \ln p = 24.7$ kcal/mol at 1873 K [3]. Values of $C_p(T)$ and S_{298} as estimated from that of Th_3N_4, ThN, and Cr_2N (extrapolated from 800 K) [7, 8] are given in Table 11 and used to calculate the reaction enthalpy, $\Delta H_{298} = +87.9$ kcal/mol. From these data the standard Gibbs free energy for the above reaction is given to within 1.5 kcal/mol by the equation ΔG (in cal/mol N_2) = 86800 − 33.16 T (T = 298 to 2200 K). From these results the standard enthalpy of formation of Th_2CrN_3 is calculated as $\Delta H_{f,298} = -221.0$ kcal/mol and the standard Gibbs free energy of formation can be represented to within 1.5 kcal/mol by the equation

$$\Delta G_f \text{ (in cal/mol)} = -219590 + 57.63\,T \quad (298 \text{ to } 2200 \text{ K}).$$

Table 11

Standard Molar Thermodynamic Properties of Th_2CrN_3 and Th_2MnN_3, Calculated by the Authors of this Article from Estimated Entropy and Heat Capacity Data of Analogous Compounds [7, 8].

compound	temperature range in K	C_p (in cal·mol⁻¹·K⁻¹) = $a + b \times 10^{-3} T + c \times 10^6 T^{-2}$			S_{298} in cal·mol⁻¹·K⁻¹	$\Delta H_{f,298}$ in kcal/mol
		a	b	c		
Th_2CrN_3	298 to 2173	32.96	7.658	−0.3237	37.6	−221.0
Th_2MnN_3	298 to 990	32.82	8.678	−0.2732	39.6	−215.5
	990 to 1360	35.45	5.958	−0.2357		
	1360 to 1410	33.15	8.858	−0.280		
	1410 to 2130	38.22	5.298	−0.2357		

References for 4.2.5 see p. 56

4.2.5.2 The Th–W–N System

The phase relations in this system are important for use of W as a crucible material for the nitrides at elevated temperatures. There is practically no mutual solubility between W and ThN and, a fortiori, Th_3N_4, as long as the N_2 pressure exceeds the decomposition pressure of ThN (cf. p. 20), by a factor of about 100. At lower pressures, ThN dissociates more or less rapidly and the product Th metal reacts rapidly with W above 1700°C to form a metal-rich liquid with a composition very near the eutectic given by Ackermann and Rauh as 1695°C with the atom fraction $x_W = 0.012$ [9]. The previously reported liquid formation at lower temperatures is discounted [10]. The solubility of W in liquid Th has been reported as given by the equations:

$$\log x_W = (0.056 \pm 0.138) - (3880 \pm 290)/T \text{ in the temperature range 1968 to 2200 K and}$$

$$\log x_W = 6.478 - 32.20 \times 10^3/T + 31.19 \times 10^6/T^2 \text{ in the temperature range 2200 to 2800 K,}$$

where x_W is the W atom fraction in the binary Th–W liquid. Because of the low N_2 pressures at which the liquid can exist near the eutectic temperature, the solubility of N_2 in the liquid is rather low. The solubility becomes appreciable in the liquid above 2200 K [11].

No binary W–Th [9] or ternary Th–W–N compound exist. A suggested 2273-K isotherm is illustrated in **Fig. 12**.

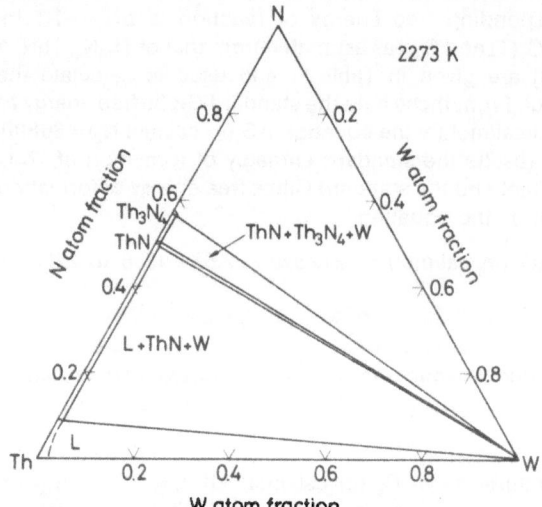

Fig. 12

Isothermal section in the Th–W–N system at 2273 K, suggested on the basis of the discussions in the text.

4.2.5.3 The Th–Mn–N System

Phase Relations

The 1573-K isotherm as illustrated in **Fig. 13** is based upon the available information on the binary Th–N [12, 13] and Mn–N [14] phase diagrams. Fig. 13 shows that Th_2MnN_3 coexists in 3-phase equilibrium with Th_3N_4 and Mn.

References for 4.2.5 see p. 56

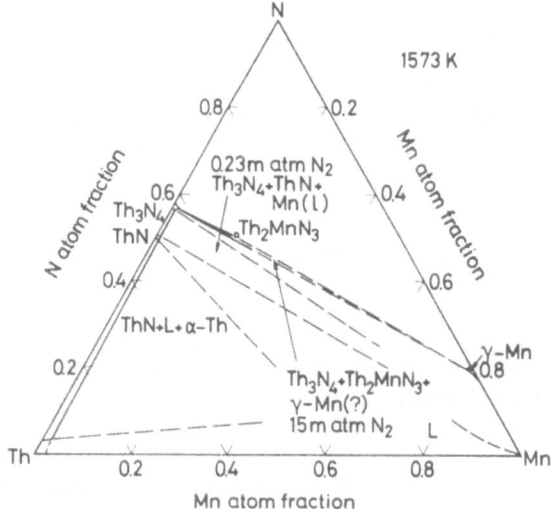

Fig. 13

Isothermal section in the Th–Mn–N system at 1573 K, postulated by the authors of this article.

Th_2MnN_3

Preparation. The compound Th_2MnN_3 is prepared by heating the powder mixture $2ThN + Mn$ in a W crucible with lid assembly in 1 atm N_2 for 2 h at 1300°C or for longer periods of time at lower temperatures [3]. For the most pure preparation, a small excess of Mn is necessary to compensate for vaporization losses.

Crystal Structure. Orthorhombic, U_2CrN_3-type, crystals of Th_2MnN_3 have the lattice parameters $a = 379.19 \pm 0.03$, $b = 354.82 \pm 0.02$, and $c = 1283.21 \pm 0.0024$ pm and the theoretical density $D_X = 10.796 \pm 0.002$ g/cm³ [3]. The structure is the same as described for Th_2CrN_3 and with the same atomic position parameters. The interatomic distances (in pm) are: Th–Th 354.82 ± 0.02 (2×); Th–N(I) 256 ± 3 (2×), Th–N(II) 259.8 ± 0.4 (4×), Th–N(II) 263 ± 8 (1×); Th–Mn 319 ± 2 (4×); N(I)–Th 256 (4×), N(II)–Th 259.8 (4×), N(II)–Th 263 (1×), Mn–N 194 ± 4 (2×), Mn–N 189.60 ± 0.02 (2×). The Th–N and Mn–N bond distances are nearly the same as those in the high coordination binary nitrides, viz. 258 pm in ThN and 179 and 198 pm in Mn_3N_2 and Mn_2N, respectively [6]. The slightly higher density and shorter Y–2N(I) distances in the Th_2XN_3 (X = Cr, Mn) compound with X = Mn than that of Cr must reflect a greater participation of the d orbitals of Mn to the bonding.

Thermodynamic Properties. The N_2 decomposition pressure of Th_2MnN_3 in the presence of excess Mn has been measured as being 11 Torr at 1300°C for the reaction [3]:

$$6\,Th_2MnN_3 \rightleftharpoons 4\,Th_3N_4 + 6\,Mn + N_2$$

for which the free enthalpy change is $\Delta G = -RT \cdot \ln p = +13.24$ kcal/mol N_2. Neglecting solubility of N_2 in Mn(l) at 1300°C and using the $C_p(T)$ and S_{298} values given in Table 11 (p. 53) as estimated from those of Th_2CrN_3 and Mn and Cr [7], the enthalpy change of the reaction in 1 atm N_2 is $\Delta H = -52.3$ kcal/mol. The caluclated free energy change of the reaction can be represented to within 0.5 kcal/mol by the equations

ΔG (in cal/mol N_2) = $53600 - 24.81\,T$ in the temperature range 298 to 1517 K and

ΔG (in cal/mol N_2) = $75300 - 39.40\,T$ in the temperature range 1517 to 2100 K.

References for 4.2.5 see p. 56

The standard enthalpy and free energy of formation are, respectively,

$\Delta H_{f, 298} = -215.5$ kcal/mol and

ΔG_f (in cal/mol) = $-213200 + 57.60$ T in the temperature range 298 to 1517 K and

ΔG_f (in cal/mol) = $-215400 + 59.10$ T in the temperature range 1517 to 2100 K.

References for 4.2.5:

[1] Mills, T. (J. Less-Common Metals **22** [1970] 373/81).
[2] Holleck, H. (Thermodyn. Nucl. Mater. Proc. 4th Symp., Vienna 1974 [1975], Vol. 2, pp. 213/63; C.A. **84** [1976] No. 5040).
[3] Benz, R., Zachariasen, W. H. (J. Nucl. Mater. **37** [1970] 109/13).
[4] Holleck, H. (KFK-1011 [1969] 1/13; N.S.A. **24** [1970] No. 3023).
[5] Holleck, H. (KFK-1726 [1972] 1/29; C.A. **79** [1973] No. 108276).
[6] Smithells, C. J. (Metals Reference Book, 6th Ed., Butterworths, London 1962, pp. 1/1087, 6/30).
[7] Kubaschewski, O., Alcock, C. B. (Metallurgical Thermochemistry, 5th Ed., Pergamon, New York 1979).
[8] Rand, M. H. (At. Energy Rev. Spec. Issue No. 5 [1975] 7/85).
[9] Ackermann, R. J., Rauh, E. G. (High Temp. Sci. **4** [1972] 272/82).
[10] Rough, F. A., Bauer, A. A. (Uranium and Thorium Alloys, Addison-Wesley, Reading, Mass. 1958; BMI-1300 [1958] 1/138; N.S.A. **12** [1958] No. 13935).

[11] Ackermann, R. J., Rauh, E. G. (High Temp. Sci. **4** [1972] 496/505).
[12] Benz, R. (J. Nucl. Mater. **31** [1969] 93/8).
[13] Benz, R., Hoffmann, C. G., Rupert, G. N. (J. Am. Chem. Soc. **89** [1967] 191/7).
[14] Hansen, M., Anderko, K. (Constitution of Binary Alloys, McGraw-Hill, New York 1958).

4.3 Thorium Hydride Nitrides

Benz, R., Naoumidis, A.
Kernforschungsanlage Jülich
Jülich, Federal Republic of Germany

4.3.1 Phase Relations in the Th–N–H System

The partial 673-K isotherm based upon published results of investigations is illustrated in **Fig. 14** [1]. A ternary compound that is assumed to be described by the compositions $ThN_{(4/3-x/3)}H_x$, where x ranges from 1 to about 2 or more, is shown, but deficiencies in the H content appear to occur at temperatures above ca. 673 K [2 to 4].

Peterson, Nelson [4] have shown that Th-rich Th + ThN alloys in H_2 atmospheres lead to the formation of ThH_2 and thereby establish the existence of the Th + ThN + ThH_2 3-phase region indicated in Fig. 14. When the Th phase has been exhausted, further addition of H leads to an increase in the H_2 equilibrium pressure to a new plateau corresponding to the coexistence of ThN, ThH_2 and the ternary compound $ThN_{(4/3-x/3)}H_x$, i.e., the ThN + ThH_2 + $ThN_{(4/3-x/3)}H_x$ phase region indicated in Fig. 14. The balanced reaction for the decomposition of this ternary compound in the 3-phase region is

$$\frac{6}{(x+2)}ThN_{(4/3-x/3)}H_x \rightleftharpoons \frac{2(4-x)}{(x+2)}ThN + \frac{2(x-1)}{(x+2)}ThH_2 + H_2.$$

The value of x at each temperature up to 600°C is unique as has been demonstrated by Peterson, Nelson [4] and there is one degree of freedom, viz. x(T) and for a specified temperature in the 3-phase region x is fixed. The composition at 400°C is estimated from the published tensimetric data as $ThN_{0.692}H_{1.924}$ [4]. The corresponding reaction is

$$1.529\,ThN_{0.692}H_{1.924} \rightleftharpoons 1.058\,ThN + 0.471\,ThH_2 + H_2 \quad\ldots\ldots\ldots\ldots\ldots\ldots\ldots\ldots\ldots\ldots\ldots (12)$$

The equilibrium pressure is given by the equation

$$\log p \text{ (p in atm)} = 3.6192 - 3570/T \quad (T = 663 \text{ to } 873 \text{ K})$$

and the free energy change is

$$\Delta G \text{ (in cal/mol } H_2) = 16340 - 16.56\,T \quad (T = 663 \text{ to } 873 \text{ K})$$

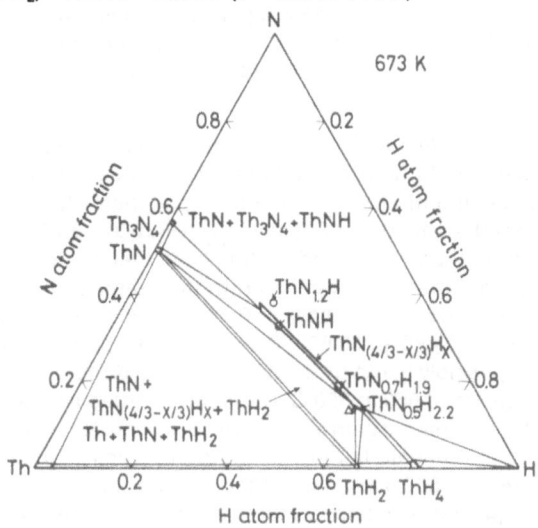

Fig. 14

Isothermal section in the Th–N–H system at 673 K [1].

The free energy change at 673 K from this equation is $\Delta G(673\text{ K}) = 5.19$ kcal/mol H_2. The heat capacity and entropy of ThH_2 are estimated from that of CeH_2, Ce [5], and Th. With these and other calorimetric data of the elements [6] and the Th compounds, the enthalpy change of reaction (12) is calculated as $\Delta H_{298} = 25.8$ kcal/mol H_2 [4].

References for 4.3.1:

[1] Benz, R. (J. Less-Common Metals **98** [1984] L17/L19).

[2] Blunck, H., Juza, R. (Z. Anorg. Allgem. Chem. **410** [1974] 9/20).

[3] Juza, R., Gerke, H. (Z. Anorg. Allgem. Chem. **363** [1968] 245/57).

[4] Peterson, D. T., Nelson, S. O. (J. Less-Common Metals **80** [1981] 221/6); see also Nelson, S. O. (IS-T-998 [1982] 1/80; INIS Atomindex 14 [1983] No. 760976).

[5] Kubaschewski, O., Alcock, C. B. (Metallurgical Thermochemistry, 5th Ed., Pergamon, New York 1979).

[6] Elliott, J. F., Gleiser, M. (Thermochemistry for Steelmaking, Vol. 1, Addison-Wesley, Reading, Mass., 1960).

4.3.2 Thorium Hydride Nitrides, ThNH, ThN$_{1.2}$N, ThN$_{(4/3-x/3)}$H$_x$

Preparation

The fcc compound of the approximate composition ThNH with a = 560 pm as first reported by Juza, Gerke [1] was obtained in an approximately equimolar mixture with the amorphous compound Th$_2$N$_2$(NH), see p. 60, when the metal was reacted with 4900 atm NH$_3$ at 320°C for 90 h in a steel vessel impervious to H$_2$. It is only a minor component in products of reaction at 500 to 550°C.

A phase evidently with the same structure and having the lattice parameter a = 516.4 pm and chemical composition ThN$_{1.2}$H was later identified in mixtures with Th(NH)$_2$ formed by reaction of Th metal powder in the presence of a small amount of NaNH$_2$ (presumably as catalyst) with 6000 atm NH$_3$ at 325°C for 13 d [2]. It was also identified in a mixture with ThH$_2$ formed by reaction with 5000 atm NH$_3$ at 200°C for 7 d. The compound was obtained pure with the composition ThN$_{1.23}$H by reaction of Th powder in the presence of NH$_4$I with 3500 atm NH$_3$ for 10 d at 300°C [2].

Peterson, Nelson [3] identified an fcc lattice phase with a = 559.6 ± 0.8 pm in the products of H$_2$ tensimetric measurements made on 2 different Th + ThN alloys. The two alloys which had been prepared by arc melting of Th in selected N$_2$ atmospheres had a much lower N$_2$ partial pressure estimated from the equation ΔG_f (cal/mol) = −88 000 + 20.24 T (T = 298 to 2027 K) as 10^{-58} atm N$_2$ at 300°C than those employed in the phase preparation by Juza and coworkers [1, 2], ca. 200 atm N$_2$ partial pressure. Compared to the preparations described by Juza and coworkers [1, 2], the N/Th ratios of the alloys of Peterson, Nelson [3] were lower by a factor of about 2; specifically, N/Th = 0.486 and 0.692. The reported H$_2$ pressure plateaus obtained with the latter alloy indicate possibly 2, but probably 3 condensed phases. The plateaus disappear at the elevated temperatures of 650 to 850°C, above which there appears to be an appreciable loss of H and probably decomposition of the ThNH structure. At the lower temperatures approaching 400°C the H contents fall in the range of values H/Th = 1.8 to 2.2. The above-quoted lattice parameter, a = 559.6 pm, was obtained with this alloy, which must have been mainly ThNH. The reported low-temperature isobars indicate its composition to be ThN$_{0.692}$H$_{1.92}$. The second alloy with the composition given as ThN$_{0.486}$H$_{2.2}$ is stated to be 3-phase, viz. ThN + ThNH + ThH$_2$. Allowing for some uncertainty in the establishment of equilibria, all the reported compositions are broadly represented by the formula ThN$_{(4/3-x/3)}$H$_x$, where the composition variable x ranges from about 1 to 2 or more [4]. In this formula, the compound is assumed to be ionic with the normal ion valences of Th^{4+}, N^{3-} and H$^-$. The composition variable x is determined by the degree of exchange between N^{3-} and H$^-$ ions on the anion sublattice. The range of compositions as given by the formula ThN$_{(4/3-x/3)}$H$_x$ is illustrated in Fig. 14. The formula is seen to be compatible with all the compositions reported from both the laboratories, that of Juza and coworkers [1, 2] and of Peterson, Nelson [3], when allowance is made for reasonable uncertainties in identification of (possible weakly crystallized) phases by X-ray diffraction techniques. A version of some possible phase relations is suggested in Fig. 14, p. 57 [4].

Properties

The X-ray diffraction pattern of ThN$_{1.23}$H indicates an fcc lattice with the parameter a = 561.4 pm which for 4 formula units gives the density D$_x$ = 9.39 g/cm^3 [2]. Pycnometric density D(25°C) = 8.95 g/cm^3 [2].

The phase of the approximate composition ThNH has the same structure with the lattice parameter a = 560 pm [1]. The fcc phase of Peterson, Nelson [3] has a = 559.6 pm, see text.

Shohoji [5] has reported results of a statistical thermodynamic analysis of bonding in fcc ThN_yH_x (y = 0.486 and 0.692) and in hcp $ThC_{0.46}H_x$. The compounds were assumed to form an ordered lattice with the N and C atoms distributed randomly over their respective octahedral and the H over tetrahedral coordinated sites. On the basis of the assumption that the Th–H bond energy in ThN_yH_x is the same as that in ThH_2, viz. $E^*_{H-Th} = -73.55$ kJ/mol, the N–H interaction energy was calculated as being $E^{ot}_{N-H} = +29.1$ to $+28.7$ kJ/mol (repulsive) in ThN_yH_x (y = 0.486 to 0.692, respectively) and that of C–H as $E^{ot}_{C-H} = -394$ kJ/mol in ThC_yH_x. The bonding and, therefore, the electronic surroundings of N in THN_yH_x are concluded to be quite different from that of the C atoms in ThC_yH_x [5].

The fcc ThNH structure has been suggested as being ionic; see "Preparation", p. 58.

The standard enthalpy of formation of $ThN_{(4/3-x/3)}H_x$ (in kcal/mol) is approximated from that of $ThN_{0.692}H_{1.924}$ by assuming constant molar heats of mixing of ThH_2, Th, and ThN as $\Delta H_{f,298}(ThN_{(4/3-x/3)}H_x) = -115.1 + 13.7x$, ΔH_f in kcal/mol, from which the standard enthalpy of formation of stoichiometric ThNH is $\Delta H_{f,298}(ThNH) = -101.4$ kcal/mol, and of $ThN_{0.692}H_{1.92}$ is -88.7 kcal/mol. The free energy of formation of ThNH thus calculated is represented by the equation [4]:

$$\Delta G_f \text{ (in cal/mol)} = -101500 + 37.1\,T \quad (T = 298 \text{ to } 900 \text{ K}).$$

For further thermodynamic data see Table 12 and "Phase Relations", p. 57. The brownish yellow compound ThNH is stable in air and dissolves slowly in warm 4N HCl with evolution of gas assumed to be H_2 [1]. Yellow $ThN_{1.2}H$ crystals are more resistant to aqueous media than the binary thorium nitrides, and Th–N–H compounds [2].

Table 12

Estimated Heat Capacity C_p (in cal·mol^{-1}·K^{-1}) and Standard Entropy S_{298} (in cal·mol^{-1}·K^{-1}) of $ThN_{(4/3-x/3)}H_x$, ThNH, and $ThN_{0.692}H_{1.92}$.
Heat capacities are estimated by the authors of this article from heat capacities of the analogous CeH_2 compound and Ce and Th elements.

compound	temperature range in K	heat capacity $C_p =$ $a + b \times 10^{-3}\,T + c \times 10^6\,T^{-2} + d \times 10^{-6}\,T^2$				S_{298}
		a	b	c	d	
$ThN_{(4/3-x/3)}H_x$	298 to 1000	13.246	2.018	−0.153	−0.485x	14.01
		−0.513x	+1.316x	+0.038x		+0.407x
ThNH		12.73	3.33	−0.115	−0.485	14.42
$ThN_{0.692}H_{1.92}$		12.26	4.55	−0.080	−0.933	14.79

References for 4.3.2:

[1] Juza, R., Gerke, H. (Z. Anorg. Allgem. Chem. **363** [1968] 245/57).
[2] Blunck, H., Juza, R. (Z. Anorg. Allgem. Chem. **410** [1974] 9/20).
[3] Peterson, D. T., Nelson, S. O. (J. Less-Common Metals **80** [1981] 221/6); see also Nelson, S. O. (IS-T-998 [1982] 1/80; INIS Atomindex **14** [1983] No. 760976).
[4] Benz, R. (J. Less-Common Metals **98** [1984] L17/L19).
[5] Shohoji, N. (J. Nucl. Mater. **127** [1985] 88/96).

4.4　Thorium Amides and Imides. Double Salts

David Brown

Chemistry Division, Atomic Energy Research Establishment,

Harwell, England

The reported thorium amides, imides, and related double salts are listed in Table 13. There has been no recent publication dealing with thorium tetramide (see "Thorium" 1955, pp. 242/3).

Table 13

Thorium Amides and Imides.

$Th(NH_2)_4$	$ThN(NH_2)$	$KTh(NH)(NH_2)_3$
$Th(NH)(NH_2)_2$	$Th_2N_2(NH)$	$KTh(NH)_2(NH_2)$
$Th_2(NH)_3(NH_2)_2$ (?)	$Th_3N_2(NH)_3$	$K_2Th_2(NH)_3(NH_2)_4$
$Th(NH)_2$		$K_2Th_2(NH)_5$

4.4.1　Thorium Amides and Imides

Preparation

The early work on thorium tetramide, $Th(NH_2)_4$, is covered in "Thorium" 1955, pp. 242/3.

The diamideimide $Th(NH)(NH_2)_2$ is reported to precipitate when potassium amide is added to a solution of the hexanitrato complex $K_2Th(NO_3)_6$ in anhydrous liquid ammonia at $-10°C$ [1].

The amidenitride, $ThN(NH_2)$, is apparently obtained by the reaction of ThNI or ThNBr with an equivalent amount of $NaNH_2$ in liquid ammonia (10 atm pressure, 20°C) or with $NaNH_2$ and NaN_3 in supercritical ammonia (4000 atm pressure, 300°C) [2]. At higher pressure (6000 atm) and with a longer reaction period (700 h, compared to 50 h) the product of the latter reaction is the diimide $Th(NH)_2$, contaminated with the imidenitride $Th_2N_2(NH)$ [2]. The diimide has also been prepared by thermal decomposition of $Th(NH)(NH_2)_2$ at 100°C (via $Th_2(NH)_3(NH_2)_2$ at 50°C) by the presumed reactions [1]:

$$2\,Th(NH)(NH_2)_2 \xrightarrow{50°C} Th_2(NH)_3(NH_2)_2 + NH_3 \text{ and}$$
$$Th_2(NH)_3(NH_2)_2 \xrightarrow{100°C} 2\,Th(NH)_2 + NH_3$$

and by heating thorium metal in ammonia (6000 atm, 325°C, 320 h) in the presence of $NaNH_2$ (0.1 g/g Th) [2] (see also [3]).

The imidenitride $Th_2N_2(NH)$ is obtained by heating thorium metal powder in ammonia (2500 to 5500 atm) at 400 to 550°C for 24 to 96 h [2, 4] and by the thermal decomposition of $Th(NH)_2$ [2]. A second imidenitride, $Th_3N_2(NH)_3$, is reported to form when ThNI or ThNBr mixed with $NaNH_2$ is heated with KN_3 in ammonia (350°C, 6000 atm, 650 h). The product also contained $Th_2N_2(NH)$ [2].

The compound $Th(NH)_2$ has been suggested [6] as one of several possibilities to explain the reaction of ThI_3 with liquid ammonia [5].

References for 4.4 see p. 62

Properties

The available X-ray powder data are listed in Table 14.

Table 14

Unit Cell Dimensions and Densities for Thorium Amides and Imides.

compound	symmetry; space group	lattice parameters (in pm)				density (in g/cm³)	Ref.
		a	b	c	β		
$Th(NH)_2$	hexagonal R$\bar{3}$m (No. 166)	395	—	2758	—	7.01[a]	[2]
						6.97[b]	[2]
$Th_2N_2(NH)$	monoclinic	717	386	624	92.2°	9.76[a]	[4]
						9.88[b]	[4]
$Th_3N_2(NH)_3$	hexagonal	398	—	3015	—	—	[2]

[a] X-ray density; [b] experimental density.

Thorium diimide is isostructural with α-Th_3N_4. The two structures are compared in **Fig. 15** [2].

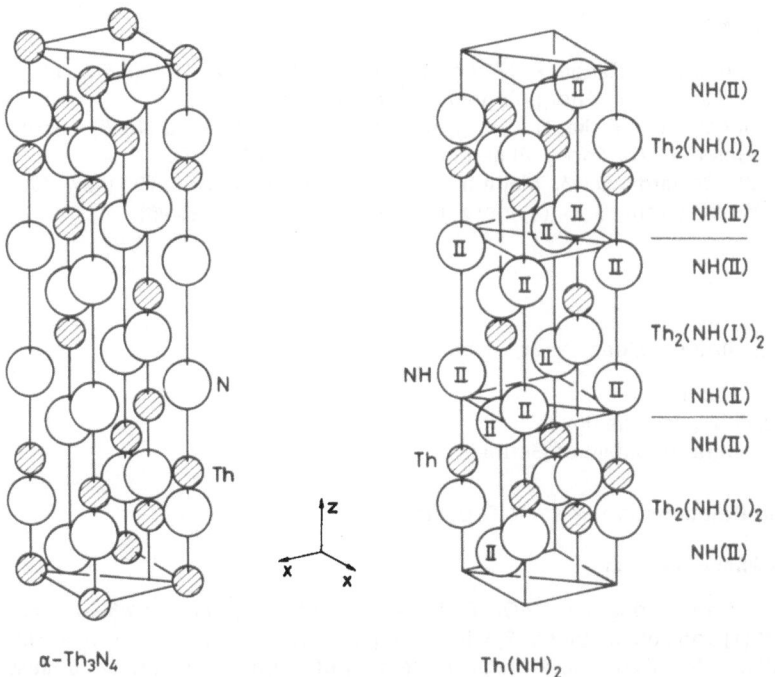

α-Th₃N₄   Th(NH)₂

Fig. 15

Comparison of the structural arrangements in α-Th_3N_4 and $Th(NH)_2$ [2].

The various thorium(IV) amides and imides are readily hydrolysed [1 to 3].

Amorphous $Th(NH)_2$ decomposes at 130°C in a stream of N_2 to amorphous Th_3N_4 [1]. In contrast, Blunck, Juza [2] found that their preparation decomposes at 100 to 330°C by the reaction $2Th(NH)_2 \rightarrow Th_2N_2(NH) + NH_3$. The yellow compound reacts explosively with water. In air it decomposes rapidly to a white substance (ThO_2) [2].

$Th(NH)(NH_2)_2$ is a white, amorphous, pyrophoric powder [1]. The thermal decomposition is mentioned above (p. 60).

$Th_2N_2(NH)$ forms greenish yellow crystals, which decompose in NH_3 at pressures as low as 0.1 atm or in vacuum to β-Th_3N_4 at 300 to 390°C [2, 4], in NH_3 at 700°C [3].

$ThN(NH_2)$ is brownish yellow [2].

4.4.2 Potassium Thorium Amideimides

Reactions involving varying ratios of KNH_2 and $K_2Th(NO_3)_6$ in liquid ammonia at −10°C are reported to yield the following complexes $KTh(NH)_2(NH_2)$, $KTh(NH)(NH_2)_3$ and $K_2Th_2(NH)_3$-$(NH_2)_4$. The last two are stable at 0°C but decompose at 20°C in a stream of N_2. The first is converted to $K_2Th_2(NH)_5$ at 150°C; this decomposes at 270°C to yield $K_3Th_3N_5$. The complexes are all white, pyrophoric solids which are readily hydrolysed [1].

References for 4.4:

[1] Schmitz-Dumont, O., Raabe, F. (Z. Anorg. Allgem. Chem. **277** [1954] 297/314).
[2] Blunck, H., Juza, R. (Z. Anorg. Allgem. Chem. **410** [1974] 9/20).
[3] Juza, R., Jacobs, H., Gerke, H. (Ber. Bunsenges. Phys. Chem. **70** [1966] 1103/5).
[4] Juza, R., Gerke, H. (Z. Anorg. Allgem. Chem. **363** [1968] 245/57).
[5] Watt, G. W., Sowards, D. M., Malhotra, S. C. (J. Am. Chem. Soc. **79** [1957] 4908/11).
[6] Watt, G. W., Malhotra, S. C. (J. Inorg. Nucl. Chem. **11** [1959] 255/6).

4.5 Thorium Nitride Oxides

Benz, R., Naoumidis, A.
Kernforschungsanlage Jülich
Jülich, Federal Republic of Germany

4.5.1 Phase Relations in the Th–N–O System

The Th–N–O Phase Diagram

Existence of a pseudobinary ThO_2–ThN eutectic with ThO_2/ThN = 15/85 was reported by Blum, Guinet [1] and substantiated [2] with the liquidus valley minimum given as ThO_2/ThN = 19/81 in 0.13 atm N_2 at 2660°C. A planar projection of the liquidus surface is illustrated in **Fig. 16** which shows sloping liquidus valleys but no distinct ternary minimum in the ThN–Th_3N_4– ThO_2 region. The minimum is too shallow for identification of its composition. The previously hypothesized monotectic in the Th–O, binary, system [3] is discounted.

References for 4.5.1 see p. 66

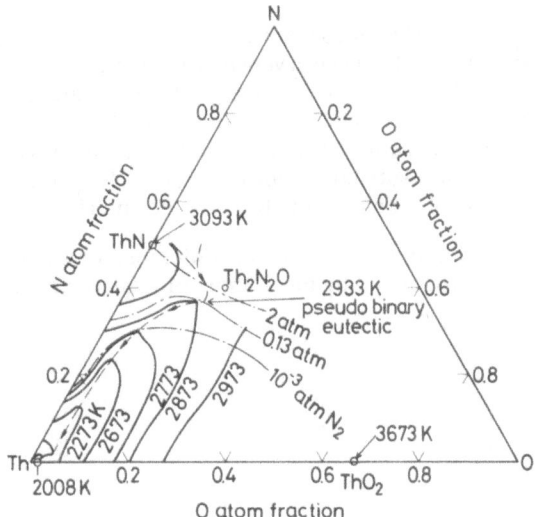

Fig. 16

Planar projection of the Th–N–O liquidus
surface [2].

An appreciable N solubility in ThO_2 at 2550 and 2660°C has been demonstrated metallographically with the atom fraction values of $x_N = 0.002$ and 0.03, respectively [2]. Incorporation of N and of Th atoms in the ThO_2 lattice must be accompanied by creation of O^{2-} ion vacancies in the nonmetal sublattice. Substoichiometry of pure ThO_2 is insignificant, $ThO_{1.998}$, above 2800 K [8].

The solubility of oxygen in ThN crystals has been demonstrated metallographically by identification of $ThO_2 +$ Th precipitates in large ThN grains after equilibrating with Th and ThO_2 at elevated temperatures [2, 4]. The solubility was determined by areal analysis of metallographic sections of the crystals after annealing for 1 d at 1200°C in vacuum to coarsen the precipitate. The results listed in Table 15 show that the precipitates appear, on the average, in equimolar amounts of Th and ThO_2 [2]. Much higher solubilities corresponding to the bulk compositions $ThN_{0.79}O_{0.19}$ (1600°C) and $ThN_{0.71}O_{0.31}$ (1800°C) have also been reported [4]. The equimolar Th + ThO_2 condition is interpreted to indicate that the ThN lattice is ordered and that for each excess Th atom located on the Th sublattice there is an associated O atom located on the nonmetal sublattice. At lower temperatures the ThO solubility is exceeded and ThO precipitates out, but, being unstable, transforms to the stable phase combination consisting of an equimolar mixture of Th + ThO_2 by the reaction $2ThO \rightleftharpoons Th + ThO_2$. These observations eliminate other conceivable lattice defect types such as anion interstitials or Th atom vacancies when ThO_2 is incorporated into the ThN lattice.

Table 15

Solubilities of Th and of ThO_2 in fcc ThN at Various Temperatures [2].

temperature in °C	2600	2500	2400	2200	2100	2000	1900	1800	1700	1600
solubility of Th in mol%	0	4	3	6	6	5	3	4	3	2
solubility of ThO_2 in mol% ..	0	3	3	4	5	3	3	3	2	2

                    References for 4.5.1 see p. 66

The ternary compound Th_2N_2O is essentially a line compound at room temperature but exists over a range of O/N ratios that fall in a very narrow section of the Th–N–O phase diagram at elevated temperatures, 1600 to 2000°C. The range of compositions can be described by the general formula $Th_2N_{2(4/3-x/3)}O_x$, where $\frac{1}{2} \leqslant x \leqslant 1$ [5]. This formula represents compositions stable with the subsequently described crystal structure built up of Th^{4+}, N^{3-}, and O^{2-} ions in which $1-x$ of the O^{2-} ions are substituted for by $\frac{2}{3}$ $(1-x)$ N^{3-} ions on the O sublattice. No appreciable substitution of O^{2-} ions for the N^{3-} ions on the N sublattice seems to occur.

Typical isotherms showing the transpiration of different 3-phase regions with temperature are illustrated in **Fig. 17** to **20**, drawn by the authors of this article on the basis of Ref. [2].

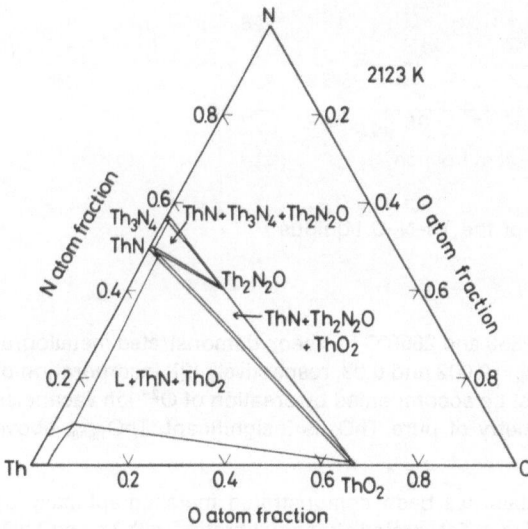

Fig. 17

Isothermal section in the Th–N–O system at 2123 K [2].

Fig. 18

Isothermal section in the Th–N–O system at 2043 K [2].

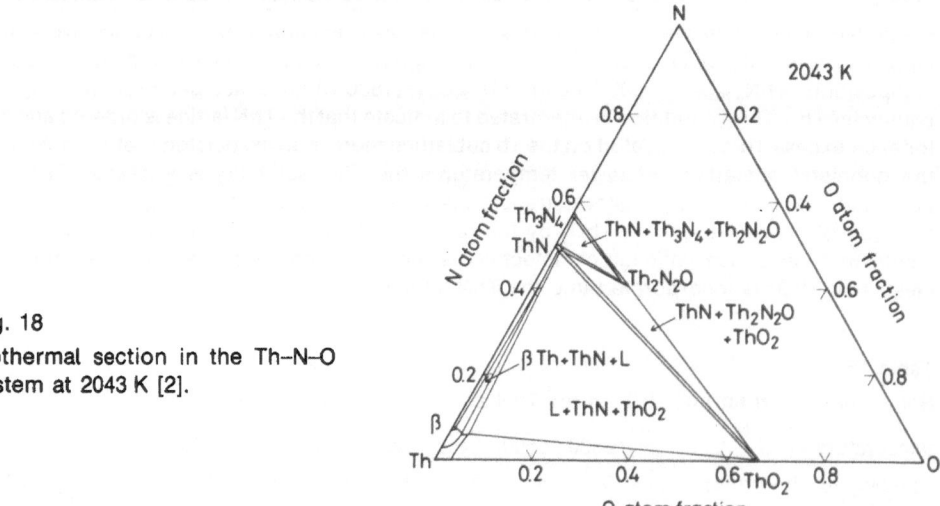

References for 4.5.1 see p. 66

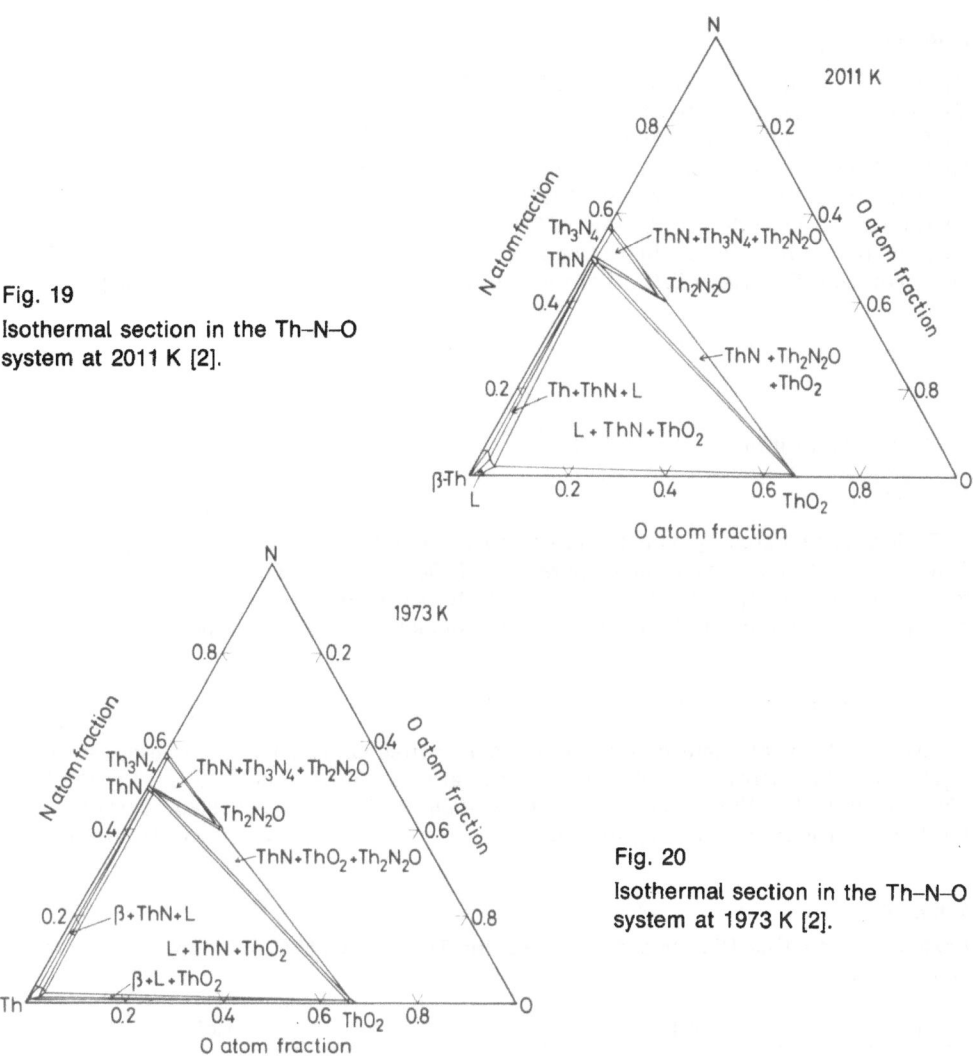

Fig. 19

Isothermal section in the Th–N–O
system at 2011 K [2].

Fig. 20

Isothermal section in the Th–N–O
system at 1973 K [2].

Influence of Nitrogen on the Electrical Resistivity of ThO$_2$

Rudolph [6] has reported the electrical conductivity of ThO$_2$ at temperatures of 1135 to
1770 K as a function of O$_2$ partial pressure. His data show an enhanced conductivity at low O$_2$
pressures, 10^{-3} to 10^{-2} atm O$_2$, corresponding to nearly pure N$_2$ which he used as a diluent
assuming the N$_2$ to be inert. With the information now available, the enhanced conductivity in
N$_2$-rich atmospheres can now be explained as due to a high sensitivity of the electrical
conductivity to change in concentration of crystal defects produced by exchange of N^{3-} for
O^{2-} ions on the anion sublattice of ThO$_2$ and the creation of a corresponding number of anion
vacancies and, possibly, electronic defects. The substitution of N^{3-} for O^{2-} ions on the ThO$_2$
lattice is sufficiently extensive in 1 atm N$_2$ at 2000°C to produce composition changes observ-
able microscopically in quench specimens [5]. High-temperature X-ray diffraction data of
Pialoux [7] seem to confirm this interpretation [7].

References for 4.5.1:

[1] Blum, P., Guinet, P. (Compt. Rend. **253** [1961] 1053/5).
[2] Benz, R. (J. Nucl. Mater. **43** [1972] 1/7).
[3] Benz, R. (J. Nucl. Mater. **29** [1969] 43/9).
[4] Kusakabe, T., Imoto, S. (Nippon Kinzoku Gakkaishi **36** [1972] 305/9).
[5] Benz, R. (J. Am. Chem. Soc. **89** [1967] 197/9).
[6] Rudolph, J. (Z. Naturforsch. **14a** [1959] 727/37).
[7] Pialoux, A. (Stud. Phys. Theor. Chem. **32** [1984] 437/44).
[8] Ackermann, R. J., Rauh, E. G., Thorn, R. J., Cannon, M. C. (J. Phys. Chem. **67** [1963] 762/9).

4.5.2 Thorium Nitride Oxide Th_2N_2O

Preparation

Th_2N_2O is prepared by equilibration of cold-pressed tablets of $3ThN + ThO_2$ [1 to 3] or $3Th + ThO_2$ [3 to 6] in 1 atm N_2 at temperatures of 1000 to 2000°C. Although the low formation temperatures of 1000 [5], 1200 [3], and 1300°C [6] have been reported, higher temperatures can be expected to improve the crystallinity and homogeneity in a shorter period of time.

Crystal Structure and Density

Reported hexagonal lattice parameters and the corresponding densities of trigonal Th_2N_2O crystals are summarized in Table 16. The crystal structure is as follows [2]: Space group $P\bar{3}m1-D^3_{3d}$ (No. 164). Hexagonal parameters: $a = 388.33 \pm 0.02$ pm, $c = 618.70 \pm 0.03$ pm. Atom positions: 2Th in $\pm (\frac{1}{3}, \frac{2}{3}, z_{Th})$, $z_{Th} = 0.250 \pm 0.006$, 2N in $\pm (\frac{1}{3}, \frac{2}{3}, z_N)$, $z_N \approx 0.631$, 1O in $(0,0,0)$.

Table 16

Hexagonal Unit Cell Dimensions Reported for Trigonal Th_2N_2O at Room Temperature.

a in pm	c in pm	D_X in g/cm³	Ref.
388.3 ± 0.2	618.8 ± 0.4	$10.44 \pm 0.02^{b)}$	[8][a)]
387	616	$10.56^{b)}$	[9][a)]
388.33 ± 0.02	618.70 ± 0.03	$10.442 \pm 0.002^{b)}$	[2]
388.3 ± 0.1	618.7 ± 0.2	$10.44 \pm 0.01^{b)}$	[1]
388.0	618.0	10.47	[10]
388.0	618.0	$10.47^{b)}$	[3]
388.0	618.0	$10.47^{b)}$	[5, 16]

[a)] Zachariasen [8] and Chiotti [9] assumed their data to be for the formula "Th_2N_3", but according to Benz, Zachariasen [2] the experimental results suggest the material described as "Th_2N_3" was Th_2N_2O. – [b)] Calculated by the authors of this article from the lattice parameters.

References for 4.5.2 see p. 69

The Th atoms are in simple hexagonal close-packing so that the c axis is twice the separation of hexagonal layers. The N atoms occupy tetrahedral and the O atoms octahedral holes of the close-packed metal structure. The atomic configuration about the Th atoms, as illustrated in **Fig. 21**, is almost identical with that of the Th(II) atoms of the related Th$_3$N$_4$ structure illustrated in Fig. 5, p. 38. Both structures are built up of hexagonal layers and both contain tetrahedrally bonded N atoms. The interatomic distances are (in pm): Th–4N 235; Th–3O 272, Th–6Th 381; Th–6Th 388. The hexagonal a axis and the layer separation are slightly smaller in Th$_3$N$_4$ [2].

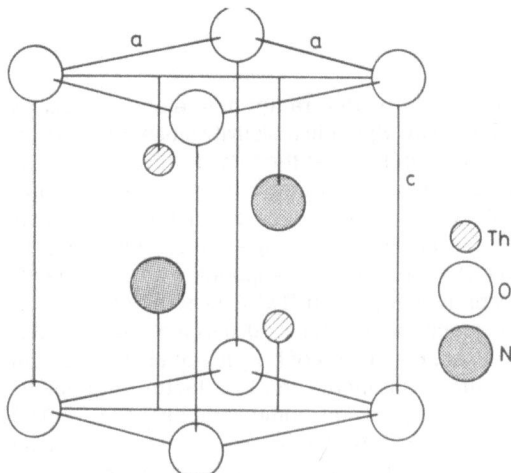

Fig. 21

Atomic configuration in the hexagonal unit cell of trigonal Th$_2$N$_2$O, drawn by the authors of this article.

Thermodynamic Properties

The N$_2$ equilibrium pressures as a function of temperature for the reaction $4\,Th_2N_2O \rightleftharpoons 2\,ThO_2 + 6\,ThN + N_2$ have been determined by an emissivity change technique and are given by the equation [5, 7]

$$\log p \text{ (p in atm)} = 13.544 - 30950/T \quad (T = 1960 \text{ to } 2260 \text{ K})$$

corresponding to the free energy of reaction

$$\Delta G \text{ (in cal/mol N}_2) = -RT \cdot \ln p = 141600 - 61.97\,T \quad (T = 1960 \text{ to } 2260 \text{ K}).$$

From these data, Kusakabe, Imoto [5, 7] estimate the standard molar enthalpy and entropy of formation at 298 K as $\Delta H_{f, 298} = -321.7$ kcal/mol and $\Delta S_{f, 298} = -76.2$ cal·mol^{-1}·K^{-1}, respectively. From the latter quantity, the absolute entropy is calculated as $S_{298}(Th_2N_2O) = 19.59$ cal·mol^{-1}·K^{-1}. This value is 34% lower than the value $S_{298}(Th_2N_2O) = 29.7 \pm 1.5$ suggested by Rand [11] and adopted here. The Th$_2$N$_2$O heat capacity as estimated by Rand [11] is

$$C_p \text{ (in cal·mol}^{-1}\cdot K^{-1}) = 27.94 + 4.48 \times 10^{-3}\,T - 0.376 \times 10^6\,T^{-2}$$

in the temperature range 298 to 2200 K. Similar calculations by the authors of this article result in

$$C_p \text{ (in cal·mol}^{-1}\cdot K^{-1}) = 27.945 + 4.236 \times 10^{-3}\,T - 0.3763 \times 10^6\,T^{-2}.$$

References for 4.5.2 see p. 69

With these estimates, ΔG of the reaction above and $\Delta H_{f,298}(ThN) = -88.5$ kcal/mol, the enthalpy of formation of Th_2N_2O is calculated using the thermal properties of Th [12] and of ThO_2 [11], O_2 and N_2 [13] as $\Delta H_f = -303.0$ kcal/mol. This value is lower in absolute magnitude than the previously given ones -321.7 [5, 7], -308 [11], and -322 kcal/mol [15]. The free energy of formation as calculated by the authors of this article from the same data by 3rd Law calculations is represented by the equation

$$\Delta G_f \text{ (in cal/mol)} = -301\,800 + 62.17\,T \quad (T = 298 \text{ to } 2200 \text{ K}).$$

Electrical Resistivity

Th_2N_2O is a semiconductor with a rather high electrical resistivity suggesting a possible appreciable ionic conductivity. Specific resistivity measurements have been made by a 4-point method as a function of N_2 pressure at the temperatures 1323, 1373, and 1423 K by Kamegashira et al. [4]. The specimen was formed by cold-pressing into a cylindrical shape and sintered for 10 h at 1350°C to 63% theoretical density. Data as read from a small-scale plot after being extrapolated 50 degrees give for the specific conductivity of 63-% dense Th_2N_2O in 1 atm N_2 the value σ (1000°C) = 0.19 S/m, corresponding to the specific resistivity R (1000°C) = 5.25 $\Omega \cdot$ m. The total conductivity of Th_2N_2O is less than that of undoped ThO_2 [14] (fired at 2200°C and with an unspecified density) at 1000°C. According to Bauerle [14], the conductivity of ThO_2 at 1000°C is mixed, consisting of (1) a major electronic component (hole conduction) $\sigma_e \approx 6$ S/m at low pressures and increasing by a factor of about 10 with increasing pressure from 10^{-5} to 1 atm O_2, and (2) a minor ionic component, $\sigma_{ion} = 0.003$ S/m, independent of O_2 pressure. The conductivity of Th_2N_2O, however, increases with decreasing N_2 pressure in the pressure range 10^2 to $10^5 Pa = 10^{-3}$ to 1 atm. Slopes of plots of log σ vs. log p_{N_2} are given as $-1/9.08$, $-1/7.63$, and $-1/7.81$ at the respective temperatures 1323, 1373, and 1423 K. The predominant lattice defects in Th_2N_2O are suggested to be triply ionized N vacancies and the conductivity n-type electronic semiconductor [4].

Magnetic Susceptibility

A temperature-independent magnetic susceptibility of 0.200×10^{-6} cm³/g (77 to 300 K) has been reported. A rather large ferromagnetic component, possibly due to an impurity, was reported [16].

Spectral Emissivity

Values reported for a pressed disk of Th_2N_2O are listed in Table 17 [5, 7]. The data must be used with reservation because the surface roughness and the gas atmosphere conditions are not specified.

Chemical Properties

The reaction $Th_2N_2O + 3H_2O \rightarrow 2ThO_2 + 2NH_3$ takes place with no gas evolution and at lower rates than for Th_3N_4 and ThN [6].

Table 17

Spectral Emissivity of Th_2N_2O at $\lambda = 0.65$ μm, $E_{0.65}$, as a Function of Temperature. The Surface Temperature is Denoted T_S [7].

T in K	T_S in K	$E_{0.65}$
2198	1999	0.367
2088	1898	0.348
1998	1823	0.346
1940	1773	0.343
1850	1777	0.331 or 0.333 [5]

References for 4.5.2:

[1] Benz, R. (J. Am. Chem. Soc. **89** [1967] 197/9).
[2] Benz, R., Zachariasen, W. H. (Acta Cryst. **21** [1966] 838/40).
[3] Kusakabe, T., Imoto, S. (Nippon Kinzoku Gakkaishi **35** [1971] 795/800; C.A. **75** [1971] No. 113503).
[4] Kamegashira, N., Tsuji, T., Miyamoto, T., Naito, K. (J. Nucl. Mater. **102** [1981] 26/9).
[5] Kusakabe, T., Imoto, S. (Technol. Rept. Osaka Univ. **22** [1972] 477/88).
[6] Sugihara, S., Imoto, S. (J. Nucl. Sci. Technol. [Tokyo] **8** [1971] 630/6).
[7] Kusakabe, T., Imoto, S. (Nippon Kinzoku Gakkaishi **35** [1971] 1115/20).
[8] Zachariasen, W. H. (Acta Cryst. **2** [1949] 388/90).
[9] Chiotti, P. (J. Am. Ceram. Soc. **35** [1952] 123/30).
[10] Juza, R., Gerke, H. (Z. Anorg. Allgem. Chem. **363** [1968] 245/57).

[11] Rand, M. H. (At. Energy Rev. Spec. Issue No. 5 [1975] 7/85, 74, 76).
[12] Hultgren, R., Desai, P. D., Hawkins, D. T., Gleiser, M., Kelley, K. K., Wagman, D. D. (Selected Values of the Thermodynamic Properties of the Elements, Metals Park, OH, 1973, pp. 1/636).
[13] Elliott, J. F., Gleiser, M. (Thermochemistry for Steelmaking, Vol. 1, Addison-Wesley, Reading, Mass., 1960).
[14] Bauerle, J. E. (J. Chem. Phys. **45** [1966] 4162/6).
[15] Wagman, D. D., Evans, W. H., Parker, V. B., Schumm, R. H., Nuttall, R. L. (NBS-TN-270-8 [1981] 1/153; C.A. **69** [1982] No. 12232).
[16] Adachi, H., Imoto, S. (Technol. Rept. Osaka Univ. **23** [1973] 1121/54, [Eng.] 425/9).

4.6 Thorium Nitrates

David Brown
Chemistry Division, Atomic Energy Research Establishment,
Harwell, England

Introduction

Thorium tetranitrate hydrates and oxide nitrate hydrates reported to-date are listed in Table 18 with the anhydrous compounds and the presently known peroxide nitrate hydrate and hydroxide nitrate hydrates. The pre-1950 literature is discussed in "Thorium" 1955, pp. 243/52. There have been no recent reports concerning the existence of $Th(NO_3)_4 \cdot 12H_2O$ and, as discussed later (Section 4.6.4, p. 75) the existence of certain other hydrates, $Th(NO_3)_4$ $\cdot xH_2O$ ($x = 6$, 2 and 1) appears to be doubtful. Thus, the best characterized tetranitrate hydrates are $Th(NO_3)_4 \cdot 5H_2O$ and $Th(NO_3)_4 \cdot 4H_2O$, and there is some evidence that $Th(NO_3)_4 \cdot 3H_2O$ can be obtained by thermal dehydration of these compounds.

Table 18

Reported Thorium Nitrates, Oxide-, Hydroxide-, and Peroxide-Nitrates.

$Th(NO_3)_4$	$ThO(NO_3)_2$	$Th_2(OH)_2(NO_3)_6 \cdot 8H_2O$	$Th_6(O_2)_{10}(NO_3)_4 \cdot 10H_2O$
$Th(NO_3)_4 \cdot xH_2O$ ($x = 12$, 6 to 1)*)	$ThO(NO_3)_2 \cdot xH_2O$ ($x = 2$, 1, 0.5)	$Th(OH)_2(NO_3)_2 \cdot 3H_2O$	
$Th(NO_3)_4 \cdot 2HNO_3 \cdot 3H_2O$	$ThO(OH)(NO_3)$?	$Th(OH)_3(NO_3) \cdot xH_2O$ ($x = 2$, 1.5)	

*) The existence of compounds with $x = 12$, 6, 2 and 1 is doubtful; see p. 75.

4.6.1 Thorium Tetranitrate $Th(NO_3)_4$

Preparation

The interaction of $Th(NO_3)_4 \cdot 4H_2O$ and an excess of dinitrogen pentoxide in anhydrous nitric acid yields $Th(NO_3)_4 \cdot 2N_2O_5$ ($= (NO_2)_2Th(NO_3)_6$, cf. p. 109) which is converted to $Th(NO_3)_4$ when heated in a vacuum at 150 to 160°C [1]. An alternative route involves the thermal decomposition of $Th(NO_3)_4 \cdot 2N_2O_4$ at 90°C in a vacuum [2]. Attempts to dehydrate $Th(NO_3)_4$ $\cdot 4H_2O$ in gaseous dinitrogen pentoxide were unsuccessful [1], and there was no evidence of reaction between either thorium metal or thorium dioxide and liquid dinitrogen tetroxide under pressure at 87°C [1]. However, interaction of $ThCl_4$ and dinitrogen tetroxide in ethyl acetate yields the complex $Th(NO_3)_4 \cdot 2N_2O_4$ which is converted to $Th(NO_3)_4$ at 150°C in a vacuum [8]. More recently electrochemical oxidation of thorium metal in a dinitrogen tetroxide-ethyl acetate-acetonitrile mixture has yielded a solution from which $Th(NO_3)_4$ was isolated [3].

The reported conversion of $Th(NO_3)_4 \cdot 4H_2O$ to $Th(NO_3)_4$ by refluxing it with ethyl acetate and allowing the product to crystallise at room temperature over a period of weeks [4] is unlikely to be correct.

Physical Properties

The structure of $Th(NO_3)_4$, which is a white solid, is unknown. The melting point is reported to be ca. 55°C [3]. The enthalpy of solution of anhydrous thorium tetranitrate at a dilution of $Th(NO_3)_4 \cdot 2500H_2O$ is reported to be -34.7 kcal/mol [5]. After making allowance for the effect of hydrolysis on the enthalpy of formation of Th^{4+} in such a solution the recommended enthalpy of formation of crystalline $Th(NO_3)_4$, $\Delta H_{f,298}(Th(NO_3)_4, c)$, is -345.5 ± 3.0 kcal/mol [6, 7].

The infrared spectrum recorded by Ferraro, Walker [8] for anhydrous thorium tetranitrate is compared in Table 19 with those reported at the same time for the tetra- and penta-hydrate (see also [3] for information on $Th(NO_3)_4$). The ultraviolet spectrum of a diethyl ether solution of the anhydrous tetranitrate (illustrated in [1]) is quite different from those of the hydrates.

References for 4.6.1 see p. 72

Table 19

Infrared Data (in cm^{-1}) for $Th(NO_3)_4$, $Th(NO_3)_4 \cdot 4H_2O$ and $Th(NO_3)_4 \cdot 5H_2O$ (s = strong, vs = very strong, m = medium, w = weak, vw = very weak, br = broad, sh = shoulder) [8].

	$Th(NO_3)_4$	$Th(NO_3)_4 \cdot 5H_2O$	$Th(NO_3)_4 \cdot 4H_2O$
H_2O stretch	3518 (s) 3440 (s) 3080 (s, br)	3518 (s) 3440 (s) 3075 (s, br)	
$\nu_1 + \nu_2$	2378 (m), 2360 (sh)	2550 (w)	
$3\nu_3$ or $\nu_4 + \nu_2$		2330 (w)	
$2\nu_2$		2075 (vw), 2060 (vw), 2040 (vw)	
$\nu_2 + \nu_3$ or $\nu_2 + \nu_5$, ν_6		1790 (vw), 1775 (vw), 1733 (vw)	
H_2O bend		1655 (s), 1630 (sh)	1660 (m), 1655 (m)
ν_1	1620 (s) 1560 (s, br)	1550 (sh) 1510 (vs, br) 1420 (sh)	1550 (sh) 1500 (s, br)
ν_4	1328 (s) 1240 (s, br)	1350 (s) 1325 (s) 1293 (vs)	1320 (m) 1290 (s)
ν_2	1010 (s, br)	1040 (sh) 1030 (s) 875 (sh)	1040 (sh) 1030 (s)
ν_3	800 (s)	815 (m), 808 (vs)	812 (s), 805 (s)
ν_5, ν_6	740 (s) 575 (vw)	760 (sh), 745 (s)	759 (s), 743 (s)
H_2O vibrations for the pentahydrate and the tetrahydrate	500 (m) 430 (vw) 350 (vw) 310 (vw)	553 (m) 474 (m) 395 (m) 365 (m)	551 (sh) 475 (m) 395 (sh) 367 (w, sh)

References for 4.6.1 see p. 72

Table 19 (continued)

	Th(NO$_3$)$_4$	Th(NO$_3$)$_4 \cdot 5$H$_2$O	Th(NO$_3$)$_4 \cdot 4$H$_2$O
coordinated nitrate (Th–O st)	244(s)	244(s)	249(m)
		211(s)	218(vw)
NO$_2$ restricted rotations Th–O–N bend lattice vibrations		178(s)	
		139(w)	139(w)
		120(vw)	121(vw)
		107(vw), 98(vw)	107(vw), 99(vw)

Chemical Properties

The chemical properties of thorium tetranitrate have not been extensively investigated.

It is readily soluble in water but dissolution in diethyl ether is more difficult [1]. It is soluble in dinitrogen tetroxide-ethyl acetate-acetonitrile mixture and addition of the appropriate ligand yields the following donor complexes, Th(NO$_3$)$_4 \cdot 8$dmso, Th(NO$_3$)$_4 \cdot 2$tppo, Th(NO$_3$)$_4$ $\cdot 8$pyNO, Th(NO$_3$)$_4 \cdot 2$bpy and Th(NO$_3$)$_4 \cdot 2$phen (dmso = dimethylsulfoxide; tppo = triphenylphosphine oxide; pyNO = pyridine-N-oxide; bpy = 2,2'-bipyridine; phen = 1,10-phenanthroline) [3].

Addition of dimethylsulfoxide to a solution produced by electrochemical oxidation of thorium in nitric acid-tributylphosphate yields Th(NO$_3$)$_4 \cdot 6$dmso whilst addition of 2,2'-bipyridine or 1,10-phenanthroline in ethanol yields, respectively, [(bpyH)$_3$NO$_3$][Th(NO$_3$)$_6$] and [(phenH)$_3$NO$_3$][Th(NO$_3$)$_6$]. The hexanitrato complex [N(C$_2$H$_5$)$_4$]$_2$[Th(NO$_3$)$_6$] is obtained by addition of a solution of tetraethylammonium nitrate in ethanol [3].

References for 4.6.1:

[1] Ferraro, J. R., Katzin, L. I., Gibson, G. (J. Am. Chem. Soc. 77 [1955] 327/9).
[2] Schmeiser, M., Koehler, G. (Angew. Chem. 77 [1955] 456/6).
[3] Kumar, N., Tuck, G. (Can. J. Chem. 62 [1984] 1701/4).
[4] Zhou, D.-Y. (KexueTongbao 24 [1979] 978/9; C.A. 92 [1980] No. 68745).
[5] Ferraro, J. R., Katzin, L. I., Gibson, G. (J. Inorg. Nucl. Chem. 2 [1956] 118/24).
[6] Cordfunke, E. H. P., O'Hare, P. A. G. (The Chemical Thermodynamic Properties of Actinide Elements and Compounds, Pt. 3, Miscellaneous Actinide Compounds, IAEA, Vienna 1978).
[7] Rand, M. (in: Kubaschewski, O., Thorium, Physico-Chemical Properties of Its Compounds and Alloys, IAEA, Vienna 1975, pp. 7/85).
[8] Ferraro, J. R., Walker, A. (J. Chem. Phys. 45 [1966] 550/3).

4.6.2 The Th(NO$_3$)$_4$–H$_2$O System

Templeton [1] studied the Th(NO$_3$)$_4$–H$_2$O system over the temperature range 20 to 160°C and reported the formation of Th(NO$_3$)$_4 \cdot 5.5$H$_2$O below 122°C and Th(NO$_3$)$_4 \cdot 4$H$_2$O above this temperature.

References for 4.6.2 see p. 74

Marshall et al. [2] reported the crystallisation of $Th(NO_3)_4 \cdot 6H_2O$ and $Th(NO_3)_4 \cdot 4H_2O$ in this system, the temperature of transition being 111.3°C. Their results are shown in Table 20 and **Fig. 22**. A lower hydrate, $Th(NO_3)_4 \cdot \dot{x}H_2O$, was postulated above 151°C.

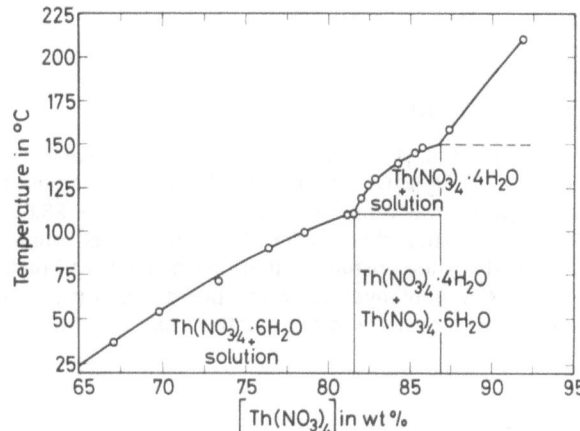

Fig. 22

The Th(NO₃)₄–H₂O system above 20°C [2].

Table 20

The Solubility of Thorium Tetranitrate in Water, 37 to 211°C [2].

temperature in °C	liquid phase wt% Th(NO₃)₄	solid phase wt% Th(NO₃)₄	solid phase
37.3	67.07	81.18 (theor., 81.62)	
54.5	69.78		
72.0	73.39		
90.2	76.39	80.82 (theor., 81.62)	$Th(NO_3)_4 \cdot 6H_2O$
99.7	78.56		
110.4	81.11		
110.9	81.50		
111[a]			$Th(NO_3)_4 \cdot 6H_2O + Th(NO_3)_4 \cdot 4H_2O$
120.6	82.01		
128	82.41		
129.5		85.84 (theor., 86.95)	
130.5	82.85		$Th(NO_3)_4 \cdot 4H_2O$
139.5	84.27		
146.0	85.30		
149.0	85.81		
151[b]			$Th(NO_3)_4 \cdot 4H_2O + Th(NO_3)_4 \cdot xH_2O$
159	87.41		$Th(NO_3)_4 \cdot xH_2O$
211	91.82		

[a] Intersection temperature for melting point of the hexahydrate. – [b] Intersection temperature for melting point of the tetrahydrate.

References for 4.6.2 see p. 74

References for 4.6.2:

[1] Templeton, C. C. (AECU-1721 [1950] 1/9; N.S.A. **6** [1952] No. 82).
[2] Marshall, W. L., Gill, J. S., Secoy, C. H. (J. Am. Chem. Soc. **73** [1951] 4991/2).

4.6.3   The $Th(NO_3)_4$–HNO_3–H_2O System

Ferraro et al. [1] showed that only the penta- and tetra-hydrate could be obtained from the $Th(NO_3)_4$–HNO_3–H_2O system (**Fig. 23**). The composition for co-existence of the penta- and tetra-hydrate was reported as 17.70 wt% water − 28.67 wt% thorium tetranitrate and that for the tetrahydrate and an unidentified "anhydrous" phase 5.10 wt% water − 21.13 wt% thorium tetranitrate. They postulated that phases such as $Th(NO_3)_6 \cdot 6H_2O$ and $Th(NO_3)_4 \cdot 5.5H_2O$ were actually the pentahydrate; X-ray powder studies confirmed [2] that $Th(NO_3)_4 \cdot 5.5H_2O$ [3] exhibited the same diffraction pattern as $Th(NO_3)_4 \cdot 5H_2O$ [2, 4].

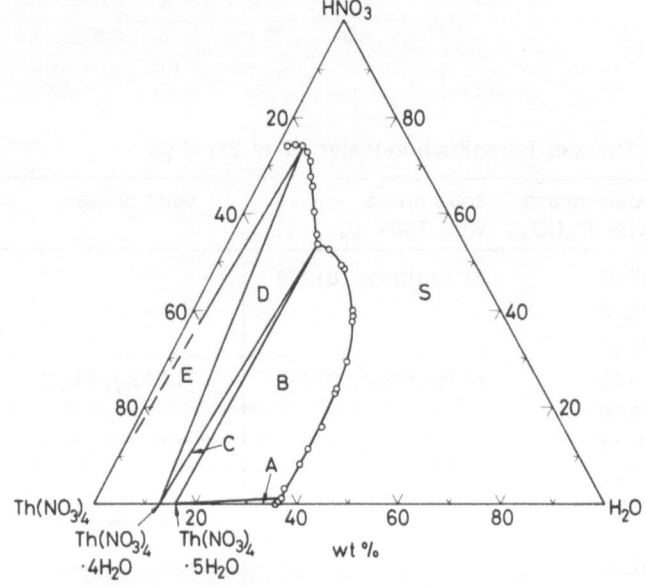

Fig. 23
The system $Th(NO_3)_4$–HNO_3–H_2O at 25°C [1].
A = region of hydrolysis; B = pentahydrate and variable liquid; C = pentahydrate, tetrahydrate and liquid; D = tetrahydrate and variable liquid; E = tetrahydrate, anhydrous phase and liquid; S = homogeneous solution.

References for 4.6.3:

[1] Ferraro, J. R., Katzin, L. I., Gibson, G. (J. Am. Chem. Soc. **76** [1954] 909/10).
[2] Staritzky, E., Walker, D. I. (LA-1439 [1952] 1/43; N.S.A. **10** [1956] No. 2391).
[3] Templeton, C. C. (AECU-1721 [1950] 1/9; N.S.A. **6** [1952] No. 82).
[4] Staritzky, E. (Anal. Chem. **28** [1956] 2021).

4.6.4 Thorium Tetranitrate Hydrates

Several thorium tetranitrate hydrates are described in the literature, Th(NO$_3$)$_4$·xH$_2$O (x=12, 6, 5.5, 4, 3, 2, 1). Those most readily obtained by crystallisation from aqueous media are Th(NO$_3$)$_4$·5H$_2$O and Th(NO$_3$)$_4$·4H$_2$O. The pre-1950 accounts of the preparation of the various phases are discussed in "Thorium" 1955, pp. 243/4 (see also [1 to 3]). The present discussion is confined to more recent publications and, for convenience, the different hydrates are dealt with in a single section.

4.6.4.1 Preparation

Two hydrates exist as solid phases in the systems Th(NO$_3$)$_4$–H$_2$O and Th(NO$_3$)$_4$–HNO$_3$–H$_2$O: the tetrahydrate and the pentahydrate. The two phases Th(NO$_3$)$_4$·6H$_2$O [4] and Th(NO$_3$)$_4$·5.5H$_2$O [7] were postulated by Ferraro et al. [5] to be the pentahydrate, see Section 4.6.3, p. 74.

An alternative route to Th(NO$_3$)$_4$·5H$_2$O involves dissolution of reagent grade thorium tetranitrate hydrate in ether followed by removal of the insoluble hydrolysed material to allow crystallisation of the pentahydrate by evaporation of the solvent [11, 12, 40].

Th(NO$_3$)$_4$·4H$_2$O is formed by crystallisation from a concentrated nitric acid solution stored over concentrated sulfuric acid in a dessicator [5]. Nikolaev et al. [8] report its formation over the concentration and acidity ranges 70 to 28.4 wt% thorium tetranitrate and 3.78 to 58.88 wt% HNO$_3$.

Although there is some evidence for the formation of lower hydrates on thermal decomposition of Th(NO$_3$)$_4$·4H$_2$O, for example, Th(NO$_3$)$_4$·3H$_2$O [9 to 12, 60, 61], Th(NO$_3$)$_4$·2H$_2$O [13] and Th(NO$_3$)$_4$·1H$_2$O [12, 13], such phases have, in general, not been well characterised and are often deficient in nitrate, e.g. [12]. However, Claudel [61] (see also [9]), in the course of an extensive investigation of the thermal dehydration of Th(NO$_3$)$_4$·5H$_2$O, obtained DTA and TGA evidence for the formation of Th(NO$_3$)$_4$·4H$_2$O at 90 to 118°C and Th(NO$_3$)$_4$·3H$_2$O at 118 to 144°C on heating at 2°C/min in air. In experiments involving heating at 5°C/day these phases were formed at 70 and 95°C, respectively. No other lower hydrates were observed during these investigations. According to [60] Th(NO$_3$)$_4$·5H$_2$O is converted to the tetrahydrate at 110°C and this decomposes to the trihydrate at 140°C.

The preparation of thorium tetranitrate hydrates on the industrial scale is described in references [14 to 17]; for earlier reports see "Thorium" 1955, pp. 244/5.

A phase of composition Th(NO$_3$)$_4$·2HNO$_3$·3H$_2$O (or, possibly, Th(NO$_3$)$_4$·(5−x)H$_2$O·xHNO$_3$) has also been reported. Thus, Moseley et al. [10] prepared Th(NO$_3$)$_4$·5H$_2$O by crystallisation from concentrated nitric acid at room temperature and dissolved this in water to give a 3.5M solution from which the new compound crystallised (typically over 5 to 7 days) as colourless needles. On heating at 55 to 60°C or on vacuum dehydration at room temperature the weight loss corresponds closely to three molecules of water; this is regained on exposure to air at room temperature. Additional information on these phases would be of value.

References for 4.6.4 see pp. 93/8

4.6.4.2   Physical Properties

Crystal Structure

Unit cell dimensions available for thorium tetranitrate hydrates are listed in Table 21. Full structural information is available for only $Th(NO_3)_4 \cdot 5H_2O$, for which both X-ray and neutron diffraction studies have been reported [18, 19]. The arrangement of the oxygen atoms around thorium as viewed in the b direction is shown in **Fig. 24** and a stereoscopic view along the Th–O(1) bond is shown in **Fig. 25**. There are eleven oxygen atoms co-ordinated to each thorium atom, eight from four bidentate groups and three from water molecules. The arrangement has two-fold symmetry along the Th–O(1) line. The hydrogen bond scheme is illustrated in **Fig. 26**. Selected bond lengths determined by the two techniques are compared in Table 22, p. 78.

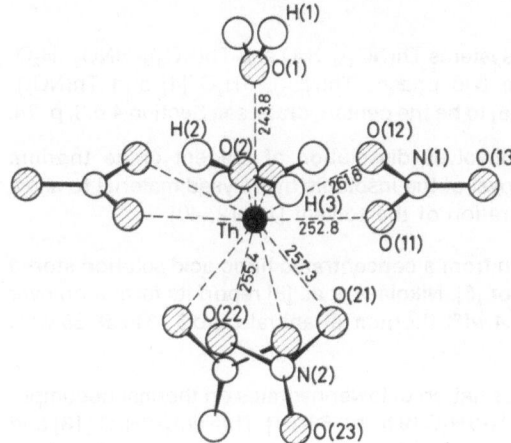

Fig. 24

Crystal structure of $Th(NO_3)_4 \cdot 5H_2O$. Oxygen arrangement around the thorium atom as viewed in the b direction [19].

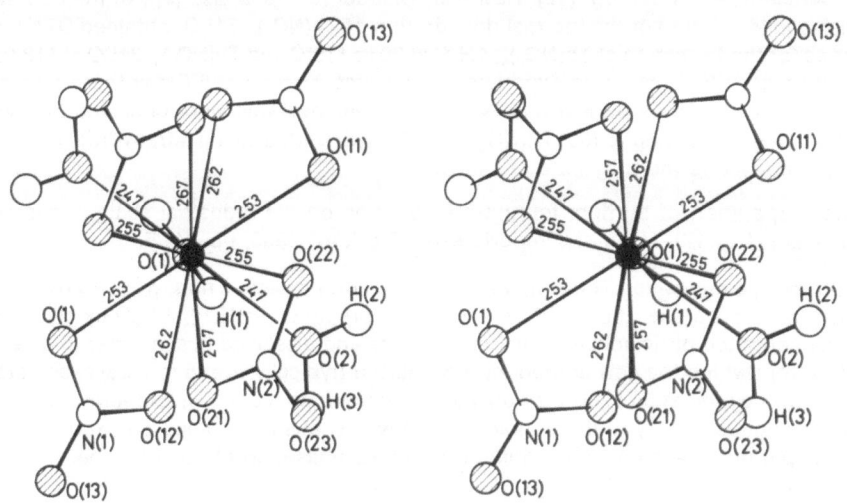

Fig. 25

Stereoscopic drawing of the oxygen arrangement around thorium in $Th(NO_3)_4 \cdot 5H_2O$ as viewed along the Th–O(1) bond [19].

References for 4.6.4 see pp. 93/8

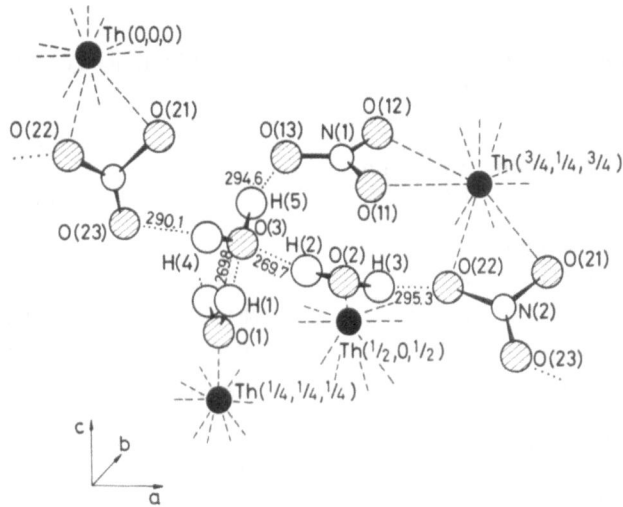

Fig. 26

Hydrogen bond scheme in Th(NO₃)₄·5H₂O [19].

Table 21

Unit Cell Dimensions for Thorium Tetranitrate Hydrates.

compound	symmetry; space group	lattice parameters in pm				D in g/cm³	Ref.
		a	b	c	β		
Th(NO₃)₄·5H₂O	orthorhombic Fdd2 (No. 43)	1118.2(3)	2287.3(5)	1057.3(3)	—	2.80[a]	[18]
						2.84[b]	[18]
Th(NO₃)·5H₂O	orthorhombic Fdd2 (No. 43)	1119.1(7)	2288.9(15)	1057.9(7)	—	—	[19]
Th(NO₃)₄·2HNO₃ ·3H₂O	monoclinic	1760(10)	1650(10)	2420(10)	88.5°(1°)	2.50[a]	[10]
						2.46[b]	[10]

[a] Calculated density. — [b] Experimental density.

X-Ray powder diffraction data for Th(NO₃)₄·4H₂O and Th(NO₃)₄·3H₂O are listed in [61].

Mechanical and Thermodynamic Properties

Calculated and experimental densities for Th(NO₃)₄·5H₂O and Th(NO₃)₄·2HNO₃·3H₂O are given in Table 21. Staritzky [20] reports an experimental value of 2.787(1) g/cm³ for the former. Densities and molar and specific volumes reported by Molodkin et al. [11] for Th(NO₃)₄·4H₂O and Th(NO₃)₄·5H₂O are included in Table 23, p. 78.

Reported enthalpies of solution for Th(NO₃)₄·4H₂O [21] and Th(NO₃)₄·5H₂O [21, 22] have been assessed by Cordfunke, O'Hare [24] (for an earlier assessment by Rand see [25]) who recommend for Th(NO₃)₄·4H₂O(c): $\Delta H^\circ_{f, 298} = -647 \pm 3.0$ kcal/mol and for Th(NO₃)₄·5H₂O(c): $\Delta H^\circ_{f, 298} = -718.9 \pm 1.0$ kcal/mol.

For additional information on the enthalpy of formation of the pentahydrate see [30].

Table 22

Comparison of Bond Lengths (in pm) in $Th(NO)_4 \cdot 5H_2O$ as Determined by X-Ray Diffraction [18] and by Neutron Diffraction [19].

	X-ray diffraction [18]	neutron diffraction [19]		X-ray diffraction [18]	neutron diffraction [19]
Th–O (nitrate)			nitrate groups		
Th–O(11)	250(1)	252.8(3)	N(1)–O(11)	127(2)	127.0(3)
Th–O(12)	262(1)	261.8(3)	N(1)–O(12)	128(2)	125.0(3)
Th–O(21)	259(1)	257.3(3)	N(1)–O(13)	121(1)	120.2(3)
Th–O(22)	258(1)	255.4(3)	N(2)–O(21)	127(2)	126.4(3)
			N(2)–O(22)	127(2)	127.5(3)
Th–O (water)			N(2)–O(23)	124(2)	120.6(3)
Th–O(1)	241(2)	243.8(5)			
Th–O(2)	248(1)	247.3(3)	hydrogen bonds		
			O(1)–O(3)	274(2)	269.8(4)
			O(2)–O(3)	271(2)	269.7(4)
			O(2)–O(22)	290(2)	295.3(4)
			O(3)–O(23)	286(2)	290.1(4)
			O(3)–O(13)	296(2)	294.6(4)

Table 23

Densities (D), Molar Volumes (V_m), Specific Volumes (V_{sp}), Refractive Indices (n_g, n_m, n_p), and Molar Refractions (R_D) of Nitrates of Thorium [11].

compound	mol mass	D in g/cm³	V_m in cm³/ mol	V_{sp} in cm³/g	n_g $(= n_\gamma)$	n_m $(= n_\beta)$	n_p $(= n_\alpha)$	n_{mean}	R_D in cm³/ mol
$Th(NO_3)_4 \cdot 4H_2O$	552.04	2.83	195.07	0.353	1.646	—	1.503	1.573	64.27
$Th(NO_3)_4 \cdot 5H_2O$	570.04	2.78	205.05	0.359	1.646	—	1.512	1.578	68.02
$NaTh(NO_3)_5 \cdot 8.5H_2O$	727.03	2.451	296.62	0.408	1.609	—	1.483	1.545	92.60
$KTh(NO_3)_5 \cdot 6H_2O$	689.14	2.844	242.31	0.351	1.545	—	1.617	1.581	80.72
$Rb_2Th(NO_3)_6$	744.98	3.42	226.60	0.292	1.618	—	1.565	1.591	81.35
$Cs_2Th(NO_3)_6$	869.94	3.642	238.86	0.274	1.617	1.599	1.581	1.610	82.80
$MgTh(NO_3)_6 \cdot 8H_2O$	772.35	2.44	316.53	0.409	1.534	1.522	1.507	1.521	96.38
$NiTh(NO_3)_6 \cdot 8H_2O$	806.75	2.54	317.61	0.393	1.549	—	1.521	1.535	98.88
$K_3Th(NO_3)_7$	783.31	2.769	282.88	0.361	1.608	—	1.577	1.592	95.70
$K_3(H_3O)_3[Th(NO_3)_{10}]$ $\cdot H_2O$	1044.34	2.274	459.25	0.439	soluble in immersion liquids				

References for 4.6.4 see pp. 93/8

The heat capacity of $Th(NO_3)_4 \cdot 5H_2O$ has been determined by low-temperature calorimetry by Cheda et al. [23]. The results and other thermal functions (5 to 350 K) are listed in Table 24. Cordfunke, O'Hare [24] recommend:

$$C^\circ_{p,298}(Th(NO_3)_4 \cdot 5H_2O(c)) = 114.9 \pm 0.1 \, cal \cdot K^{-1} \cdot mol^{-1}$$
$$S^\circ_{298}(Th(NO_3)_4 \cdot 5H_2O(c)) = 129.8 \pm 0.1 \, cal \cdot K^{-1} \cdot mol^{-1}$$

and calculate:

$$\Delta S^\circ_{f,298}(Th(NO_3)_4 \cdot 5H_2O(c)) = -547.1 \pm 0.1 \, cal \cdot K^{-1} \cdot mol^{-1}$$
$$\Delta G^\circ_{f,298}(Th(NO_3)_4 \cdot 5H_2O(c)) = -556.0 \pm 1.0 \, kcal/mol$$

All four values are virtually identical with those reported by Cheda et al. [23].

Table 24
Thermodynamic Properties of $Th(NO_3)_4 \cdot 5H_2O$ (1 cal_{th} = 4.184 J) [23].

T in K	C_p in $cal_{th} \cdot K^{-1} \cdot mol^{-1}$	$S^\circ_T - S^\circ_0$ in $cal_{th} \cdot K^{-1} \cdot mol^{-1}$	$H^\circ_T - H^\circ_0$ in cal_{th}/mol	$-(G^\circ_T - H^\circ_0)/T$ in $cal_{th} \cdot K^{-1} \cdot mol^{-1}$
5	0.136	(0.043)	(0.170)	(0.009)
10	1.044	0.352	2.654	0.086
15	2.824	1.089	12.046	0.286
20	5.163	2.211	31.819	0.620
25	7.919	3.652	64.37	1.077
30	11.003	5.364	111.56	1.645
35	14.330	7.307	174.81	2.313
40	17.80	9.446	255.10	3.069
45	21.33	11.747	352.93	3.904
50	24.86	14.177	468.41	4.809
60	31.74	19.32	751.8	6.795
70	38.13	24.70	1101.6	8.967
80	43.92	30.18	1512.4	11.276
90	49.10	35.66	1978.0	13.681
100	53.74	41.08	2492.6	16.15
110	57.96	46.40	3051.4	18.66
120	61.84	51.61	3650.7	21.19
130	65.50	56.71	4287.5	23.73
140	68.99	61.69	4960.1	26.26
150	72.37	66.57	5667	28.79
160	75.66	71.34	6407	31.30
170	78.88	76.03	7180	33.79
180	82.02	80.62	7985	36.26
190	85.08	85.14	8820	38.72
200	88.05	89.58	9686	41.15

Table 24 (continued)

T in K	C_p in $cal_{th} \cdot K^{-1} \cdot mol^{-1}$	$S_T^\circ - S_0^\circ$ in $cal_{th} \cdot K^{-1} \cdot mol^{-1}$	$H_T^\circ - H_0^\circ$ in cal_{th}/mol	$-(G_T^\circ - H_0^\circ)/T$ in $cal_{th} \cdot K^{-1} \cdot mol^{-1}$
210	90.95	93.95	10581	43.56
220	93.78	98.24	11505	45.95
230	96.56	102.47	12456	48.32
240	99.30	106.64	13436	50.66
250	102.02	110.75	14442	52.98
260	104.72	114.80	15476	55.28
270	107.41	118.81	16537	57.56
280	110.09	122.76	17624	59.82
290	112.76	126.67	18739	62.06
300	115.41	130.54	19879	64.27
310	118.05	134.37	21047	66.47
320	120.70	138.16	22240	68.65
330	123.38	141.91	23461	70.82
340	126.10	145.63	24708	72.96
350	128.81	149.33	25983	75.09
273.15	108.26	120.06	16876	58.27
298.15	114.92	129.83	19666	63.87

The enthalpies of solution of $Th(NO_3)_4 \cdot 4H_2O$ and $Th(NO_3)_4 \cdot 5H_2O$ in water and in various organic solvents as reported by Ferraro et al. [21] are listed in Table 25.

Using the solubility of ca. 3.7 mol/kg for $Th(NO_3)_4 \cdot 5H_2O$ in H_2O–HNO_3 [5] ΔG_S is calculated as -2.26 ± 0.1 kcal/mol [24]. Enthalpies of dilution of $Th(NO_3)_4 \cdot aq$ have been measured [27, 28].

Table 25

Enthalpies of Solution (ΔH_S) of Thorium Nitrate Hydrates in Water and in Various Organic Solvents at 25°C (values in kcal/mol) [21].
The numbers in parenthesis indicate the solvent/solute mole ratios.

solvent/solute	$Th(NO_3)_4 \cdot 4H_2O$	$Th(NO_3)_4 \cdot 5H_2O$
water	-7.6_5 (350)	-3.5_5 (350)
tributyl phosphite	-42.9 (480)	-44.2 (80)
		-37.2 (480)[*]
dimethyl formamide	-25.2 (150)	-21.6 (150)
		-20.0 (293)[*]
dibutyl butylphosphonate	-18.8 (450)	-14.2 (80)
		-15.5 (450)

References for 4.6.4 see pp. 93/8

Table 25 (continued)

solvent/solute	$Th(NO_3)_4 \cdot 4H_2O$	$Th(NO_3)_4 \cdot 5H_2O$
tetrahydrofuran	-14.2 (80)	-9.4_0 (80)
		-8.9_7 (138)[*]
		-5.9_5 (300)[*]
tributyl phosphate	-12.1 (80)	-7.6_5 (80)
ethylene glycol diethyl ether	-11.1 (150)	-6.6_0 (150)
diethyl ether	-9.5_5 (80)	-4.8_5 (80)
		-4.5_2 (110)[*]
		-3.25 (218)[*]
ethylene glycol monoethyl ether	-9.4_0 (150)	-7.0_5 (150)
dibutyl "carbitol"	-8.4_0 (450)	-3.3_0 (450)
acetone	-6.6_5 (80)	-3.3_5 (80)
methyl ethyl ketone	-5.9_5 (130)	-1.50 (130)
ethyl acetate	-1.41 (200)	3.10 (165)
		1.04 (200)[*]
methyl isobutyl ketone	-1.19 (300)	1.74 (100)
		0.37 (300)[*]
ethyl propionate	1.05 (150)	4.4_5 (150)
ethyl chloroacetate	1.38 (300)	4.7_0 (300)
n-amyl acetate	1.10 (200)	4.6_5 (130)
		5.0_7 (200)[*]
isobutyl alcohol	1.65 (180)	6.3_5 (180)
diethyl malonate		9.8_5 (700)

[*] Single determinations.

Magnetic Properties

As would be expected $Th(NO_3)_4 \cdot 4H_2O$ is reported to be diamagnetic [63].

Optical and Spectroscopic Properties

Staritzky [20] reports the following data for $Th(NO_3)_4 \cdot 5H_2O$: refractive indices, $n_x = 1.518$, $n_y = 1.528$, and $n_z = 1.628$; molecular refraction: 65.88 cm³/mol; optic axial angle: $2V_z = 38°$ [20]. Values for the refractive indices n_g, n_m, and n_p ($= n_\gamma$, n_β and n_α, respectively) and n_{mean} and the molar refraction R_D are given in Table 23, p. 78 [11] for the penta- and tetra-hydrates.

References for 4.6.4 see pp. 93/8

Vibrational spectral data have been reported for solid thorium tetranitrate hydrates by several investigators [31, 32, 34 to 42]. The most extensive infrared assignments for $Th(NO_3)_4 \cdot 5H_2O$ and $Th(NO_3)_4 \cdot 4H_2O$ are those reported by Ferraro, Walker [39]. Their results are listed in Table 19, p. 71, together with data on $Th(NO_3)_4$. Raman nitrate modes (in cm^{-1}) are reported by Petrov et al. [40] for the pentahydrate at 635, 714 (v_6), 740, 762 (v_3), 820 (v_5), 950, 1035 (v_2), 1230, 1323 (v_4), 1496, 1536 (v_1) and 1625 (δH_2O), whereas Ferraro et al. [38] report somewhat different positions, viz., 760, 810, 1044, 1104 and 1213 cm^{-1}. Bands for the tetrahydrate occur at 714, 748, 1036, 1487 and 1542 cm^{-1} [32].

Photoelectron (ESCA) spectra have been reported for $Th(NO_3)_4 \cdot 5H_2O$ [58], $Th(NO_3)_4 \cdot 4H_2O$ [59] and a phase of undetermined degree of hydration [57]. Weak satellites ~ 6 eV from the primary $4f_{7/2}$ and $4f_{5/2}$ peaks are assigned to ligand-to-thorium 5f shake-up transitions [58]. Spectra recorded for $Th(NO_3)_4 \cdot 4H_2O$ $(0 \rightarrow \approx 1000$ eV) are illustrated in [59] and binding energies are tabulated in [62].

4.6.4.3   Chemical Properties

Hydration, Dehydration, Stability

The thermal dehydration of $Th(NO_3)_4 \cdot 5H_2O$ and $Th(NO_3)_4 \cdot 4H_2O$ has been the subject of several investigations [9 to 13, 31, 41, 60, 61, 64 to 66] since publication of "Thorium" 1955 (see pp. 243/4). As indicated earlier (p. 75) there is more recent evidence for the formation of a trihydrate [9 to 12, 60, 61], but no other lower hydrate, and the subsequent formation of an oxide nitrate [9, 61] or oxide nitrate hydrate [41] has been suggested (Section 4.6.6, p. 99) but such phases are not well characterized. The ultimate product of thermal dehydration/denitration is thorium dioxide and there have been numerous studies into the preparation of it by this route (see, for example, references [14, 15, 67 to 77] and "Thorium" Suppl. Vol. C 1, 1978, pp. 61/7).

$Th(NO_3)_4 \cdot 5H_2O$ is stable at ambient temperature for at least two months at relative humidities in the range 55 to 60% [22, 23]; it is deliquescent at 30°C in a water vapour saturated atmosphere [61]. According to Claudel [61] when stored for several weeks over phosphorus pentoxide it is converted to $Th(NO_3)_4 \cdot 3H_2O$. The hydration of $Th(NO_3)_4 \cdot 4H_2O$, to yield the pentahydrate, is relatively slow; for example 76% conversion occurs in about 24 h at 21°C and 55% relative humidity [61]. Surprisingly, it is reported that at 18°C, 60% relative humidity rehydration of $Th(NO_3)_4 \cdot 3H_2O$, to give the tetrahydrate, is only ca. 70% complete after 23 d [61].

Solubility

The $Th(NO_3)_4$–H_2O [4, 7] and $Th(NO_3)_4$–HNO_3–H_2O [5, 8] systems are discussed in Sections 4.6.2 and 4.6.3, pp. 72/4. The data reported by Ferraro et al. [5] indicate a solubility of ca. 3.7 mol/kg for $Th(NO_3)_4 \cdot 5H_2O$; others have reported values of 3.74 mol/kg [26] and 3.66 mol/kg [22]. According to Nikolaev et al. [8] solid $Th(NO_3)_4 \cdot 4H_2O$ exists in the $Th(NO_3)_4$–$UO_2(NO_3)_2$–HNO_3–H_2O system in equilibrium with solid $UO_2(NO_3)_2 \cdot 4H_2O$ at a solution composition of 51.14 wt% Th, 10.10 wt% U, 5.07 wt% HNO_3; higher thorium concentrations result in dissolution of $UO_2(NO_3)_2 \cdot 4H_2O$.

Solubility data reported by Phillips, Huber [56] for the $Th(NO_3)_4$–$Al(NO_3)_3$–HNO_3 system at 25 and 50°C are listed in Table 26.

Some data for the solubility of $Th(NO_3)_4 \cdot 5H_2O$ and $Th(NO_3)_4 \cdot 4H_2O$ in non-aqueous media are listed in Table 27. More extensive results reported by Katzin et al. [81] on isobutyl alcohol, hexone, diethyl ether, ethylene glycol diethyl ether, ethylene glycol dibutyl ether, diethylene glycol diethyl ether, diethylene glycol dibutyl ether, tributyl phosphate, dibutyl butylphosphonate and tributyl phosphite are listed in Tables 28 and 29, pp. 84/5. These data were obtained by contacting excess solid commercial thorium nitrate hydrate with the appropriate solvent or, where necessary, the solid in contact with a saturated aqueous solution. The water correction (column 5, Table 28) is based on determination of "free" water in the organic phase and allows calculation of the H_2O:Th ratio based on the excess water.

Table 26

Thorium Nitrate Solubility in Nitric Acid-Aluminium Nitrate at 25 and 50°C [56].

HNO_3 in mol/L	$Th(NO_3)_4$ in mol/L	$Al(NO_3)_3$ in mol/L	HNO_3 in mol/L	$Th(NO_3)_4$ in mol/L	$Al(NO_3)_3$ in mol/L
25°C			50°C		
2.77	1.94	0.30	2.2	2.20	0.30
6.4	1.49	0.30	3.05	2.11	0.30
9.83	0.93	0.30	6.0	1.66	0.30
			9.6	1.38	0.30
2.6	1.74	0.60			
5.03	1.47	0.60	2.2	2.05	0.60
6.40	1.19	0.60	5.3	1.59	0.60
			8.0	1.38	0.60
2.70	1.50	0.90			
3.10	1.43	0.90	1.41	1.90	0.90
3.60	1.29	0.90	2.3	1.77	0.90
5.76	0.67	0.90	4.50	1.51	0.90
			5.62	1.35	0.90

Table 27

Solubility Date for $Th(NO_3)_4 \cdot 5H_2O$ and $Th(NO_3)_4 \cdot 4H_2O$ (in g/100 g).

solvent		$Th(NO_3)_4$ $\cdot 5H_2O$	$Th(NO_3)_4$ $\cdot 4H_2O$	Ref.
n-hexylalcohol	(pure)	33.5	33.8	[78]
methyl-n-hexylketone	(practical)	31.3	—	[78]
	(pure)	30.7	32.3	[78]
tributylphosphate		42.6/42.4	—	[79, 80]
tributylthiophosphate		2.1/1.8	—	[79]

Table 28

Composition of Organic Layers in Equilibrium with Saturated Thorium Nitrate [81].

solvent composition in %[a]	Th in %	Th(NO$_3$)$_4$ in %	water in %	water cor. in %	excess water in %	H$_2$O/Th	org./Th[b]
isobutyl alcohol							
100	23.95	49.53	15.75	1.79	13.95	7.5	4.55
71.5	18.68	38.67	11.45	1.22	10.23	7.06	6.0
50.2	13.41	27.76	7.53	0.84	6.69	6.43	7.6
43.0	11.51	23.83	6.18	0.70	5.48	6.14	8.2
	10.70	22.13	5.95	0.72	5.23	6.30	9.1
33.3	8.22	17.02	4.66	0.55	4.11	6.44	10.0
	8.14	16.83	4.42	0.55	3.87	6.12	10.1
20.0	3.88	8.03	2.03	0.30	1.73	5.75	14.5
9.0	0.78	1.61	0.42	0.10	0.32	(5.3)	—
isobutyl alcohol[c]							
100	23.6	48.8	15.2	2.06	13.1	7.15	—
89.8	20.76	42.93	13.42	1.71	11.71	7.27	—
77.7	18.40	38.05	11.00	1.37	9.63	6.74	—
71.4	16.75	36.64	10.20	1.20	9.00	6.92	—
63.8	phase separation						
hexone							
100	22.68	46.90	12.15	0.30	11.85	6.74	4.2
	22.23	45.98	11.90	0.30	11.60	6.72	4.4
83.5	18.48	38.22	9.46	0.28	9.18	6.40	5.5
71.7	15.50	32.06	7.88	0.26	7.62	6.33	6.5
60.3	11.69	24.18	6.03	0.21	5.82	6.42	8.3
	11.35	23.48	5.66	0.21	5.45	6.18	8.7
43.2	4.26	8.81	2.24	0.18	2.06	6.23	21
33.5	1.12	2.32	1.27	0.12	1.15	13.2	67
25.2	0.21	0.43	0.53	—	—	—	—
20.1	0.00	0.00	0.36	—	—	—	—
hexone[c]							
92.2	19.20	39.71	9.98	0.70	9.28	6.22	—
85.5	17.71	36.63	9.12	0.60	8.52	6.20	—
78	phase separation						
diethyl ether							
100	21.08	43.59	12.38	—[d]	—	7.58	6.5
76.7	14.75	30.50	7.75	—	—	6.78	13
53.7	3.20	6.62	1.66	—	—	6.68	20
36.3	0.54	1.12	0.38	—	—	(4.4)	—

References for 4.6.4 see pp. 93/8

Table 28 (continued)

solvent composition in %[a]	Th in %	Th(NO₃)₄ in %	water in %	water cor. in %	excess water in %	H₂O/Th	org./Th[b]
			ethylene glycol diethyl ether				
100	27.90	57.73	16.30	0.22	16.08	7.42	1.8
82.4	24.25	50.15	13.50	0.20	13.30	7.07	2.4
			phase separation				
			ethylene glycol dibutyl ether				
100	14.24	29.46	5.75	0.20	5.55	5.02	6.1
84.1	8.86	18.33	3.69	0.19	3.50	5.09	9.9
77.8	6.83	14.13	2.81	0.16	2.65	5.00	12.6
61.2	2.50	5.17	1.22	0.14	1.08	5.5	31
42.7	0.70	1.45	0.40	0.12	0.28	5.1	—
			diethylene glycol diethyl ether				
100	27.80	57.49	16.12	0.19	15.93	7.38	1.36
85.2	25.37	52.47	14.15	0.19	13.96	7.09	1.61
			phase separation				
			diethylene glycol dibutyl ether				
100	21.65	44.79	8.90	0.35	8.55	5.09	2.3
68.9	15.45	31.97	6.42	0.10	6.32	5.28	2.9
43.9	8.07	16.70	3.18	0.05	3.13	5.00	4.6
26.9	0.96	1.99	0.47	0.01	0.46	6.2	29

[a] Per cent. oxygenated component; second component, CCl_4. – [b] Oxygenated component/Th, corrected for CCl_4. – [c] n-Heptane diluent rather than CCl_4. – [d] Pure water solubility only 1.25%, so corrections less than for hexone (<0.20 at·100%).

Table 29

Composition of Organic Phosphate Layers in Equilibrium with Saturated Thorium Nitrate [81].

solvent composition in %[a]	ThO₂ in %	Th(NO₃)₄ in %	H₂O in %	H₂O/Th	P/Th
		tributyl phosphate			
100	22.92	41.67	0.58	0.37	2.50
84.6	20.88	37.96	0.49	0.34	2.47
59.2	16.24	29.53	0.45	0.40	2.53
38.4	11.90	21.64	0.38	0.47	2.50
20.8	6.83	12.41	0.15	0.32	2.64
6.60	2.18	3.96	0.04	0.3	2.89

References for 4.6.4 see pp. 93/8

Table 29 (continued)

solvent composition in %[a)	ThO$_2$ in %	Th(NO$_3$)$_4$ in %	H$_2$O in %	H$_2$O/Th	P/Th
dibutyl butylphosphonate					
100	24.79	45.07	1.00	0.59	2.30
84.4	22.43	40.78	0.79	0.51	2.32
58.2	17.63	32.05	0.63	0.52	2.35
37.5	12.81	23.29	0.38	0.44	2.36
20.5	7.84	14.25	0.24	0.45	2.36
6.28	2.72	4.94	0.13	0.7	2.33
tributyl phosphite[b)					
100	27.29	49.61	7.06	3.8	1.68
84.1	25.08	45.59	6.02	3.5	1.72
83.6	24.46	44.47	6.06	3.6	1.79
57.2	19.88	36.14	4.76	3.5	1.80
36.5	14.51	26.38	3.09	3.1	1.87
19.9	7.98	14.51	1.60	2.9	2.22
19.8	8.60	15.63	1.85	3.2	2.00
6.16	2.80	5.09	0.46	2.4	2.20

[a) Second component of solvent, CCl$_4$. – [b) H$_2$O/Th and P/Th taking water analyses at face value.

Data obtained by Verstegen [82] by contacting solid Th(NO$_3$)$_4 \cdot 5$H$_2$O with tri-n-octylamine nitrate in various diluents are listed in Table 30. Additional data on amine nitrates systems are available in reference [83].

Table 30
Solubility of Th(NO$_3$)$_4 \cdot 5$H$_2$O in Tri-n-Octylamine Nitrate Solutions, Expressed in mol/L, Calculation of the Loading Number n [82].

diluent	amine salt concentration in mol/L	Th(NO$_3$)$_4$ in mol/L	n
nitrobenzene	0.1	0.042	2.4
	0.05	0.023	2.2
	0.025	0.011	2.2
o-dichlorobenzene	0.1	0.049	2.0
	0.05	0.023	2.1
	0.025	0.014	1.8

References for 4.6.4 see pp. 93/8

Table 30 (continued)

diluent	amine salt concentration in mol/l	Th(NO$_3$)$_4$ in mol/l	n
chloroform	0.1	0.037	2.7
	0.05	0.019	2.6
	0.025	0.0094	2.7
benzene	0.1	0.0060[a]	[b]
	0.05	0.0055[a]	[b]
	0.025	0.0057	4.4
modified dodecane	0.1	0.0042[a]	[b]
	0.05	0.0045[a]	[b]
	0.025	0.0052[a]	[b]

[a] Concentration of metal nitrate in the light part of the organic phase. –
[b] Loading number not calculated, owing to the uncertainty in the trioctylamine concentration.

The solubility of Th(NO$_3$)$_4$·5H$_2$O in ether-diluent mixtures has been reported by Vdovenko et al. [84]; the results, obtained by contacting the solid nitrate with the appropriate non-aqueous phase, are shown in Table 31. Earlier information on the solubilities of thorium tetranitrate hydrates in non-aqueous media will be found in "Thorium" 1955, pp. 249/51.

Table 31

Solubility of Th(NO$_3$)$_4$·5H$_2$O in Ether-Diluent Mixtures at 25°C (c$_A$ = concentration of the solvent in mol/L, c$_M$ = concentration of Th(NO$_3$)$_4$·5H$_2$O in mol/L) [84].

c_A	c_M	c_A	c_M	c_A	c_M
benzene		butyl chloride		chlorobenzene	
1.98	0.00562	1.38	0.00085	1.70	0.00126
2.37	0.00871	1.70	0.00170	1.88	0.00178
2.88	0.0135	2.88	0.0100	3.55	0.0120
2.82	0.0186	5.31	0.0776	4.36	0.0364
4.85	0.0987	6.92	0.190	5.82	0.0832
4.85	0.105			7.25	0.120
5.37	0.126			7.66	0.229
6.24	0.234				
6.24	0.251				
7.29	0.380				

References for 4.6.4 see pp. 93/8

Table 31 (continued)

c_A	c_M	c_A	c_M	c_A	c_M
cyclohexane		CCl$_4$		chloroform	
1.57	0.00065	1.91	0.00100	2.40	0.00182
2.04	0.00190	2.24	0.00204	3.80	0.00892
2.14	0.00174	3.02	0.00562	4.57	0.0148
2.57	0.00364	4.21	0.0155	5.43	0.0316
4.79	0.0436	6.84	0.0955	6.60	0.0676
5.55	0.0760	7.95	0.195	9.12	0.263
6.45	0.158				
7.25	0.246				

Reactions

Thorium tetranitrate hydrates react with nitrogen donor ligands such as amines, diamines, pyridine and its derivatives, etc. and oxygen donor ligands such as alcohols, phenols, crown ethers, amides, heterocyclic N-oxides, phosphine oxides, sulfoxides, etc. to yield a variety of solid complexes. Examples of the different classes of compound are given in Tables 32 to 38 inclusive, together with the appropriate references. Generally the preparations involve a tetranitrate hydrate and the ligand either in a non-aqueous solvent or alone, although in a few instances complexes have been prepared by addition of the ligand to an aqueous nitric acid solution containing Th^{4+}. Recently a number of complexes have been prepared by addition of the appropriate ligand to anhydrous thorium nitrate in dinitrogen tetroxide-ethyl acetate-acetonitrile mixture (see Section 4.6.1, p. 72). The properties of these complexes are described in "Thorium" Suppl. Vol. E, 1985, pp. 1/112. This volume also provides information on complexes involving Schiff Bases and the conversion of thorium tetranitrate hydrates to a wide variety of chelate complexes.

The preparation and properties of thorium(IV) nitrato complexes are dealt with in Chapter 4.7 (p. 107).

Table 32

Thorium Tetranitrate Complexes with Nitrogen Donor Ligands.

complex	ligand L	Ref.
Th(NO$_3$)$_4$·xL·yH$_2$O	L = butylamine; x = 1, y = 3; x = 2, y = 1; x = 3, y = 4; x = 4, y = 4	[85]
	L = dimethylamine; x = 1, y = 2; x = 4, y = 8	[85]
	L = triethylamine; x = 1, y = 2; x = 2, y = 1	[85]
	L = 1,2-diaminopropane; x = 2, y = 2; x = 4, y = 1	[86]
	L = 1,4-diaminobutane; x = 2, y = 3	[86]
	L = 1,2-diaminobenzene, x = 2, y = 0	[87]
	L = pyridine, α-picoline, 2,4-lutidine and 2,6-lutidine, x = 2, y = 0	[88]

References for 4.6.4 see pp. 93/8

Table 32 (continued)

complex	ligand L	Ref.
	L = picolinic acid, x = 4, y = 0	[88]
	L = quinoline and isoquinoline, x = 2, y = 0	[88]
	L = piperidine, x = 2, y = 6	[89]
	L = 2,2'-bipyridine; x = 1, y = 0; x = 2, y = 4	[88, 89]
	L = 4,4'-dimethyl-2,2'-bipyridine, x = 1.5, y = 0	[89]
	L = 1,10-phenanthroline, x = 2, y = 0	[90]
	L = 2,9-dimethyl-1,10-phenanthroline, x = 2, y = 4	[89]
	L = 1,2-dipyridylethane, x = 1, y = 1	[91]
	L = dipyridylamine, x = 2, y = 0	[89]
	L = pyrazine, x = 2, y = 6; x = 3, y = 0	[89, 92]
	L = pyrazine-2-carboxamide, x = 2, y = 0	[92]
	L = 2-(2'-pyridyl)benzimidazole, x = 2, y = 0	[93]
	L = 2-guanidinobenzimidazole, x = 2, y = 3	[94]
	L = hexamethylmelamine (= triazine), x = 1, y = 0	[95]
	L = terpyridine, x = 1.5, y = 0	[89]
$Th(NO_3)_4 \cdot xL \cdot 4CH_3OH$	L = piperazine, N-methylpiperazine, and N,N-dimethyl-piperazine, x = 1	[96, 97]
	L = 2-methylpiperazine, x = 2	[97]
$Th(NO_3)_4 \cdot xL \cdot 2C_2H_5OH$	L = piperazine, N-methylpiperazine, and 2-methyl-piperazine, x = 2	[96, 97]
	L = N,N-dimethylpiperazine, x = 1	[97]
$Th(NO_3)_4 \cdot xL \cdot thf$	L = piperazine, 2-methylpiperazine and N,N'-dimethyl-piperazine, x = 2	[96, 97]
	L = N-methylpiperazine, x = 3	[97]
$Th(NO_3)_4 \cdot 2.5L \cdot L'$	L = N-phenylpiperazine, L' = methanol, ethanol and tetrahydrofuran (thf)	[97]

Table 33
Thorium Tetranitrate Complexes with Phenols, Amides, and Substituted Amides.

complex	ligand L	Ref.
$Th(NO_3)_4 \cdot xL$	L = 2(dimethylaminoethyl)-4-methylphenol, x = 4	[98]
	L = 2(diethylaminoethyl)-4-t-pentylphenol, x = 3	[98, 99]
	L = 2,4,6-tris(dimethylaminomethyl)phenol, x = 1	[98, 99]

References for 4.6.4 see pp. 93/8

Table 33 (continued)

complex	ligand L	Ref.
	L = 6,6'-thiobis-4-t-butyl-2-(diethylaminomethyl)-phenol, x = 1	[98, 99]
$Th(NO_3)_4 \cdot xL \cdot yH_2O$	L = urea, x = 2, y = 1; x = 2, y = 2; x = 2, y = 6; x = 3, y = 1; x = 4, y = 4; x = 5, y = 3; x = 6, y = 0; x = 6, y = 2; x = 6, y = 4; x = 7, y = 2.5; x = 8, y = 0; x = 10, y = 0; x = 11, y = 2.5	[100 to 104]
$Th(NO_3)_4 \cdot xL$	L = N-methylacetamide, x = 3	[105]
	L = N-ethylacetamide, x = 3	[106]
	L = N,N-dimethylformamide, x = 2.5	[107]
	L = N,N-dimethylacetamide, x = 2.5, 3 and 4	[108, 109]
	L = N,N-diethylacetamide, x = 2.5	[105]
	L = N,N-diisopropylacetamide, x = 2.5	[106]
	L = N,N-diphenylacetamide, x = 2	[105]
	L = N,N-dicyclohexylacetamide, x = 2	[110]
	L = N,N-dimethylpropionamide, x = 2.5	[105]
	L = N,N-diisopropylpropionamide, x = 2	[110]
	L = N,N-diethylpropionamide, x = 2.67	[110]
	L = N,N-dimethylpivalamide, x = 2 and 2.5	[105, 110]
$Th(NO_3)_4 \cdot 2L$	L = salicylidene-, vanillidene- and anisylidenesalicylic acid hydrazide	[111]
	L = salicylidene benzoic acid hydrazide	[112]
$Th(NO_3)_4 \cdot xL \cdot yH_2O$	L = antipyrine, x = 2, y = 0	[113, 114]
	L = 4-aminoantipyrine, x = 2, y = 2; x = 4, y = 4	[115]
	L = diantipyrylmethane, x = 3, y = 0	[116]

Table 34
$Th(NO_3)_4$ Crown Ether Complexes.

ligand L	complex	Ref.
15-crown-5 (=1,4,7,10,13-pentaoxacyclopentadecane)	$[Th(NO_3)_4 \cdot (H_2O)_3]_3 \cdot 5L$	[117]
benzo-15-crown-5 (= 2,3,5,6,8,9,11,12-octahydro-benzo[b]-1,4,7,10,13-pentaoxacyclopentadecin)	$[Th(NO_3)_4 \cdot (H_2O)_3]_3 \cdot 5L$	[117]
18-crown-6 (=1,4,7,10,13,16-hexaoxacyclooctadecane)	$Th(NO_3)_4 \cdot L \cdot H_2O$	[119]

References for 4.6.4 see pp. 93/8

Table 34 (continued)

ligand L	complex	Ref.
	$Th(NO_3)_4 \cdot L \cdot 3H_2O$	[117]
	$2Th(NO_3)_4 \cdot 3L \cdot 2H_2O$	[118, 120]
	$Th(NO_3)_4 \cdot L \cdot 3H_2O \cdot thf$	[119]
dibenzo-18-crown-6 (= 2,3,5,6,11,12,14,15-octa-hydrodibenzo[b.k.]-1,4,7,10,13,16-hexaoxacyclo-octadecin)	$3Th(NO_3)_4 \cdot 5L \cdot 3H_2O$	[118, 120]
dicyclohexyl-18-crown-6 (= eicosahydrodibenzo-[b.k.]-1,4,7,10,13,16-hexaoxacyclooctadecin)	$Th(NO_3)_4 \cdot L \cdot 3H_2O$	[117]
	$Th(NO_3)_4 \cdot 2L \cdot HNO_3$	[121]
dibenzo-24-crown-8 (= 2,3,5,6,8,9,14,15,17,18,20,21-dodecahydrodibenzo[b.n.]-1,4,7,10,13,16,19,22-octaoxacyclotetraeicosin)	$Th(NO_3)_4 \cdot L \cdot 3H_2O$	[117]
dicyclohexyl-24-crown-8 (= tetraeicosahydrodi-benzo[b.n.]-1,4,7,10,13,16,19,22-octaoxacyclo-tetraeicosin)	$Th(NO_3)_4 \cdot 2L \cdot HNO_3$	[121]
kryptofix ⟨2,2,2⟩ (= 4,7,13,16,21,24-hexaoxa-1,10-diazobicyclo[8,8,8]hexacosane)	$Th(NO_3)_4 \cdot L \cdot 4H_2O$	[122]
2,4-diketo-16-crown-5 (=1,4,7,10,13-pentaoxa-cyclohexadecane-2,4-dione)	$Th(NO_3)_4 \cdot L \cdot H_2O$	[123]

Table 35

Thorium Tetranitrate Complexes with Pyridine N-oxides.

complex	ligand L	Ref.
$Th(NO_3)_4 \cdot xL$	L = pyridine-N-oxide, x = 8 (also x = 2 with one molecule of ethylacetate)	[89, 124]
	L = 2-methylpyridine-N-oxide, x = 3	[124]
	L = 2,6-dimethylpyridine-N-oxide, x = 3 and 4	[124, 126]
	L = 2,4,6-trimethylpyridine-N-oxide, x = 3	[124]
	L = 3-t-butyl-2,4,6-trimethylpyridine-N-oxide, x = ?	[127]
	L = quinoline-N-oxide, x = 3	[124]
	L = 2,2'-bipyridine-N-oxide, x = 2	[128]
	L = 2,2'-bipyridine-N,N'-oxide, x = 1	[129]
	L = 1,10-phenanthroline-N-oxide, x = 2	[128]
	L = 1,10-phenanthroline-N,N'-oxide, x = 2	[130]

                    References for 4.6.4 see pp. 93/8

Table 36

Thorium Tetranitrate Complexes with Sulfoxides.

complex	ligand L	Ref.
$Th(NO_3)_4 \cdot x L \cdot y H_2O$	L = dimethylsulfoxide, x = 3, y = 0 and 2; x = 4, y = 0; x = 6, y = 0; x = 7, y = 2; x = 8, y = 2; x = 9, y = 0; x = 11, y = 1; x = 12, y = 0	[131 to 136]
$Th(NO_3)_4 \cdot x L$	L = diethylsulfoxide, x = 3	[137 to 139]
	L = di-n-butylsulfoxide, x = 2	[140]
	L = di-n-pentylsulfoxide, x = 2	[140]
	L = di-n-hexylsulfoxide, x = 2	[137, 140]
	L = di-n-heptylsulfoxide, x = 2	[140]
	L = di-n-octylsulfoxide, x = 2 and 3	[140, 141]
	L = diphenylsulfoxide, x = 3 and 4	[90, 139, 142 to 144]
	L = dinaphthylsulfoxide, x = 3	[139]
	L = tetramethylenesulfoxide, x = 6	[125, 145]

Table 37

Thorium Tetranitrate Complexes with Phosphine Oxides.

complex	ligand L	Ref.
$Th(NO_3)_4 \cdot x L$	L = trimethylphosphine oxide, x = 2.33, 2.67, 3, 3.67, 4 and 5	[146, 147]
	L = tri-n-propylphosphine oxide, x = 2.67	[146]
	L = tri-i-propylphosphine oxide, x = 2	[110]
	L = tri-n-butylphosphine oxide, x = 2 and 4	[146, 148]
	L = tri-i-butylphosphine oxide, x = 3	[110]
	L = tri-n-octylphosphine oxide, x = 2 and 3	[148]
	L = hexamethylphosphoramide, x = 2, 2.67, 3 and 4	[146, 149 to 151]
	L = triphenylphosphine oxide, x = 2	[89, 146, 152]
	L = tribenzylphosphine oxide, x = 2.67	[110]
	L = methyldiphenylphosphine oxide, x = 2.67 and 3	[110]
	L = dimethylphenylphosphine oxide, x = 4	[110]
	L = bis(dicyclohexylphosphinyl)methane, x = 1	[153, 154]
	L = bis(diethylphosphinyl)ethane, x = 3	[29]
	L = octamethyldiphosphoramide, x = 1.5 and 2	[146]
	L = bis(diphenylphosphino)methane, x = 1.5	[146]
	L = bis(diphenylphosphino)ethane, x = 1.5 and 2	[33, 146]

Table 37 (continued)

complex	ligand L	Ref.
$Th(NO_3)_4 \cdot xL$	L = (2-nitro)phenyl-bis-phenylphosphine oxide, x = 2	[43]
	L = triferrocenylphosphine oxide, x = 2	[43]
$2Th(NO_3)_4 \cdot 3L$	L = bis(di-2-ethylbutylphosphinyl)methane	[154]

Table 38

Thorium Tetranitrate Complexes with Phosphate Esters and Phosphonate Esters.

ligand L	R	complex	Ref.
$(RO)_3PO$	$R = CH_3$	$Th(NO_3)_4 \cdot 3L$	[44]
		$Th(NO_3)_4 \cdot 4L$	[44]
	$R = n\text{-}C_4H_9$	$Th(NO_3)_4 \cdot 2L$	[45 to 52]
		$Th(NO_3)_4 \cdot 2.35L$	[52, 146]
	$R = i\text{-}C_4H_9$	$Th(NO_3)_4 \cdot 3L$	[53]
	$R = C_6H_5$	$Th(NO_3)_4 \cdot 4L$	[54]
$(RO)R'PO$	$R = R' = n\text{-}C_4H_{10}$	$Th(NO_3)_4 \cdot xL$ (x = 2.21, 2.67)	[55]

References for 4.6.4:

[1] Katzin, L. I. (Proc. 1st Intern. Conf. Peaceful Uses At. Energy, Geneva 1955, Vol. 7, pp. 407/13).

[2] Flahaut, J. (in: Pascal, P., Nouveau Traite De Chimie Minérale, Masson, Paris 1963, pp. 124/30).

[3] Katzin, L. I. (in: Seaborg, G. T., Katz, J. J., The Actinide Elements, McGraw-Hill, New York – Toronto – London 1954, Natl. Nucl. Energy Ser. Div. IV A **14** [1954] 81/2).

[4] Marshall, W. L., Gill, J. S., Secoy, C. H. (J. Am. Chem. Soc. **73** [1951] 4991/2).

[5] Ferraro, J. R., Katzin, L. I., Gibson, G. (J. Am. Chem. Soc. **76** [1954] 909/10).

[6] Staritzky, E., Walker, D. I. (LA-1439 [1952] 1/43; N.S.A. **10** [1956] No. 2391).

[7] Templeton, C. C. (AECU-1721 [1950] 1/9; N.S.A. **6** [1952] No. 82).

[8] Nikolaev, A. V., Ryagabinin, A. I., Afanas'ev, Yu. A. (Izv. Sibirsk. Otd. Nauk SSSR Ser. Khim. Nauk **1966** 129/31; C.A. **65** [1966] 8071).

[9] Bussière, P., Claudel, B., Renouf, J. P., Trambouze, Y., Prettre, M. (J. Chim. Phys. **58** [1961] 668/74).

[10] Moseley, P. T., Sanderson, S. W., Wheeler, V. J. (J. Inorg. Nucl. Chem. **33** [1971] 3975/6).

[11] Molodkin, A. K., Belyakova, Z. V., Ivanova, O. M. (Zh. Neorgan. Khim. **16** [1971] 1582/9; Russ. J. Inorg. Chem. **16** [1971] 835/9).

[12] Vdovenko, V. N., Statesvich, V. P., Suglobov, D. N. (Radiokhimiya **8** [1966] 211/8; Soviet Radiochem. **8** [1966] 194/9).

[13] Vdovenko, V. M., Statesvich, V. P., Suglobova, I. G., Suglobov, D. N. (Radiokhimiya **10** [1968] 6/12; Soviet Radiochem. **10** [1968] 4/9).

[14] Fareeduddin, S., Garg, R. K., Sethna, H. M. (Proc. 2nd Intern. Conf. Peaceful Uses At. Energy, Geneva 1958, Vol. 4, pp. 208/14).

[15] Braun, C., Lorrain, C., Mahut, R., Marriette, R., Muller, S., Prugnard, S. (Proc. 2nd Intern. Conf. Peaceful Uses At. Energy, Geneva 1958, Vol. 4, pp. 202/7).

[16] Albonico, S. M., Zimmerman, H. F. (Anales. Asoc. Quim. Arg. **52** [1964] 35/43; C.A. **63** [1965] 11059).

[17] Ipekeghu, B., Acarkan, S. (Turk. J. Nucl. Sci. **11** [1984] 167/70; C.A. **102** [1985] No. 187382).

[18] Ueki, T., Zalkin, A., Templeman, D. H. (Acta Cryst. **20** [1966] 836/41).

[19] Taylor, J. C., Mueller, M. H., Hitterman, R. L. (Acta Cryst. **20** [1966] 842/51).

[20] Staritzky, E. (Anal. Chem. **28** [1956] 2021).

[21] Ferraro, J. R., Katzin, L. I., Gibson, G. (J. Inorg. Nucl. Chem. **2** [1956] 118/24).

[22] Morss, L. R., McCue, M. C. (J. Chem. Eng. Data **21** [1976] 337/41).

[23] Cheda, J. A. R., Westrum Jr., E. F., Morss, L. R. (J. Chem. Thermodyn. **8** [1976] 25/9).

[24] Cordfunke, E. H. P., O'Hare, P. A. G. (The Chemical Thermodynamic Properties of Actinide Elements and Compounds, Pt. 3, Miscellaneous Actinide Compounds, IAEA, Vienna 1978).

[25] Rand, M. (in: Kubaschewski, O., Thorium, Physico-Chemical Properties of its Compounds and Alloys, IAEA, Vienna 1975, pp. 7/85).

[26] Apelblat, A., Azouloy, D., Sahar, A. (J. Chem. Soc. Faraday Trans. I **69** [1973] 1618/23).

[27] Lange, E., Miedever, W. (Z. Elektrochem. **61** [1957] 407/9).

[28] Apelblat, A., Sahar, A. (J. Chem. Soc. Faraday Trans. I **71** [1975] 1667/70).

[29] Lane, B. C. (Diss. Oxford Univ. 1970).

[30] Devina, O. A., Medvedev, V. A. (Probl. Kalorim. Khim. Termodin. Dokl. 10th Vses Konf., Moscow 1984, Vol. 1, pp. 146/8; C.A. **102** [1985] No. 138645).

[31] Cho, J. S., Wadsworth, M. E. (TID-16465 [1961] 1/13; N.S.A. **16** [1962] No. 28924).

[32] Mathien, J. P., Lounsbury, M. (Discussions Faraday Soc. No. 9 [1950] 196/207).

[33] Wassef, M. A., Hegazi, W. S., Ali, S. A. (Commun. Fac. Sci. Univ. Ankara B **27** [1981] 225/37; C.A. **97** [1982] No. 229051).

[34] Ferraro, J. R. (J. Inorg. Nucl. Chem. **10** [1959] 319/22).

[35] Miller, F. A., Carlson, G. L., Bentley, F. F., Jones, W. H. (Spectrochim. Acta **16** [1960] 135/235).

[36] Ferraro, J. R. (J. Mol. Spectrosc. **4** [1960] 99/105).

[37] Addison, C. C., Gatehouse, B. M. (J. Chem. Soc. **1960** 613/6).

[38] Ferraro, J. R., Ziomek, J. S., Mack, G. (Spectrochim. Acta **17** [1961] 802/14).

[39] Ferraro, J. R., Walker, A. (J. Chem. Phys. **45** [1966] 550/3).

[40] Petrov, K. I., Molodkin, A. K., Saralidze, O. D., Belyakova, Z. V. (Zh. Neorgan. Khim. **12** [1967] 2974/82; Russ. J. Inorg. Chem. **12** [1967] 1573/7).

[41] Vdovenko, V. M., Statesvich, V. P., Suglobov, D. N. (Radiokhimiya **8** [1966] 211/8; Soviet Radiochem. **8** [1966] 194/9).

[42] Deptula, A. (Nukleonika **17** [1972] 47/58).

[43] Casellato, U., Fregona, D., Tamburini, S., Vigato, P. A., Graziani, R. (Inorg. Chim. Acta **110** [1985] 41/6).

[44] Legin, E. K., Malinka, G. A., Shpunt, L. B., Shcherbakova, L. L. (Radiokhimiya **22** [1980] 300/3).

[45] Healy, T. V., McKay, H. A. C. (Trans. Faraday Soc. **52** [1956] 633/42).

[46] Nadig, E. W., Smutz, M. (IS-595 [1963] 1/108; C.A. **60** [1964] 5054).

[47] Saito, N., Yamaski, A. (Bull. Chem. Soc. Japan **36** [1963] 1055).

[48] Mills, A. L., Logan, W. R. (J. Inorg. Nucl. Chem. **26** [1964] 2191/3).
[49] Ferraro, J. R. (J. Inorg. Nucl. Chem. **10** [1959] 319/22).
[50] Bagnall, K. W., Wakerley, M. W. (J. Less-Common Metals **37** [1974] 149/55).

[51] Ferraro, J. R., Cristallini, C., Fox, I. (J. Inorg. Nucl. Chem. **29** [1967] 139/48).
[52] Apelblat, A., Hornick, A. (Israel J. Chem. **7** [1969] 45/8).
[53] Horx, M., Rost, J. (Z. Chem. [Leipzig] **6** [1966] 33/4).
[54] Apelblat, A., Levin, R. (J. Inorg. Nucl. Chem. **41** [1979] 115/6).
[55] Katzin, L. I., Ferraro, J. R., Wendlandt, W. W., McBeth, R. L. (J. Am. Chem. Soc. **78** [1956] 5139/44).
[56] Phillips, J. F., Huber, H. D. (BNWL-240 [1968] 1/36; N.S.A. **22** [1968] No. 40151).
[57] Allen, G. C., Tucker, P. M. (Chem. Phys. Letters **43** [1976] 254/7).
[58] Bancroft, G., Sham, T. K., Esquivel, J. L., Larsson, S. (Chem. Phys. Letters **51** [1977] 105/10).
[59] Teterin, Yu. A., Babaev, A. S., Gagarin, S. G., Klimov, V. D. (Radiokhimiya **27** [1985] 3/13; Soviet Radiochem. **27** [1985] 1/9).
[60] Tetner, R., Zaborenko, K. B. (Radiokhimiya **8** [1966] 485/9; Soviet Radiochem. **8** [1966] 451/4).

[61] Claudel, B. (Diss. Univ. Lyon 1962, pp. 1/95).
[62] Schofield, J. H. (J. Electron Spectrosc. Relat. Phenom. **8** [1976] 129/37).
[63] Belova, V. I., Syrkin, Ya. K., Molodkin, A. K., Ivanova, O. M., Shiporina, L. M. (Zh. Neorgan. Khim. **13** [1968] 1458/60; Russ. J. Inorg. Chem. **13** [1968] 766/7).
[64] Claudel, B., Trambouze, Y. (Compt. Rend. **253** [1961] 2950/2).
[65] Tiwari, R. N., Sinha, D. N. (Indian Chem. J. **14** [1980] 25/8).
[66] Wendlandt, W. W. (Anal. Chim. Acta. **15** [1956] 435/9).
[67] List, H. L. (CF-58-9-35 [1958] 1/24; N.S.A. **19** [1959] No. 2703).
[68] Nelson, E. N. (MCW-1511 [1966] 1/20; N.S.A. **21** [1967] No. 5995).
[69] Robertson, W. J., Kerr, G. E. (U.S. 3376116 [1967] 1/4).
[70] Harada, Y., Baskin, Y., Handwork, J. H. (J. Am. Ceram. Soc. **45** [1962] 253/7).

[71] Hayashi, H., Watanabe, T. (Nippon Kagaku Kaishi **1973** 858/60).
[72] Pope, J. M., Radford, K. C. (J. Nucl. Mater. **52** [1974] 251/4).
[73] Moorehead, D. R., McCartney, E. R. (J. Australian Ceram. Soc. **12** [1976] 27/33).
[74] Toshiba Corp. (Japan. 82-149829 [1982] 1/3; C.A. **98** [1983] No. 80363).
[75] Burke, T. J. (WAPD-TM-1524 [1982] 1/19; INIS Atomindex **15** [1984] No. 010286).
[76] Yahia, M., Mamoon, E. L., Bishay, A. F. (Powder Met. Intern. **17** [1985] 32/3).
[77] Bachelard, R., Joubert, P. (Fr. 143726 [1984] 1/24).
[78] Templeton, D. C. (AECU 1722 [1950] 1/11; N.S.A. **6** [1952] No. 83).
[79] Wendlandt, W. W., Bryant, J. M. (J. Phys. Chem. **60** [1956] 1145/6).
[80] Wendlandt, W. W., Bryant, J. M. (Science **123** [1956] 1121/2).

[81] Katzin, L. I., Ferraro, J. R., Wendlandt, W. W., McBeth, R. L. (J. Am. Chem. Soc. **78** [1956] 5139/44).
[82] Verstegen, J. M. P. J. (J. Inorg. Nucl. Chem. **26** [1964] 1589/99).
[83] Mezhov, E. A., Sokolov, V. S., Shesterikov, V. N., Schmidt, V. S. (Radiokhimiya **13** [1971] 456/9; Soviet Radiochem. **13** [1971] 475/8).
[84] Vdovenko, V. M., Statsevich, V. P., Suglobov, D. N. (Radiokhimiya **14** [1972] 136/40; Soviet Radiochem. **14** [1972] 138/41).
[85] Molodkin, A. K., Kharitonov, Yu. Ya., Nefedov, V. I., Belyakova, V. Z. (Zh. Neorgan. Khim. **21** [1976] 2465/70; Russ. J. Inorg. Chem. **21** [1976] 1355/7).

[86] Molodkin, A. K., Ivanova, O. M., Balakaeva, T. A., Belyakova, Z. V. (Zh. Neorgan. Khim. 25 [1986] 158/62; Russ. J. Inorg. Chem. 25 [1980] 84/6).

[87] Patil, B. K., Gaikwad, D. M. (Marathwada Univ. J. Sci. 17 [1979] 23/6; C.A. 93 [1980] No. 160399).

[88] Agarwal, R. K., Srivastava, A. K., Srivastava, M., Bhakru, N., Srivastava, T. N. (J. Inorg. Nucl. Chem. 42 [1980] 1775/8).

[89] Wassef, M. A., Hegazi, S., Ali, A. (Commun. Fac. Sci. Univ. Ankara B 27 [1981] 225/37).

[90] Smith, B. C., Wassef, M. A. (J. Chem. Soc. A 1968 1817/8).

[91] Seminara, A., Giuffrida, S., Bruno, G., Condorelli, G. (Chim. Ind. [Milan] 58 [1976] 657/8).

[92] Jain, S. C., Gill, M. S., Rae, G. S. (J. Indian Chem. Soc. 58 [1981] 210/3).

[93] Dash, K. C., Mohanta, H. (J. Inorg. Nucl. Chem. 40 [1978] 499/501).

[94] Hussain, M. S., Ali, T., Ali, S. M. (Pakistan J. Sci. Ind. Res. 16 [1973] 96/9).

[95] Gunduz, N. (Commun. Fac. Sci. Univ. Ankara B 19 [1972] 109/13).

[96] Manhas, B. S., Trikha, A. K., Singh, M. (Indian J. Chem. A 20 [1981] 196/7).

[97] Manhas, B., Trikha, A., Singh, M. (J. Inorg. Nucl. Chem. 43 [1981] 305/7).

[98] Garnovskii, A. D., Osipov, O. A., Ismailov, Kh. M., Chikina, M. L. (Zh. Neorgan. Khim. 12 [1967] 159/62; Russ. J. Inorg. Chem. 12 [1967] 80/2).

[99] Ismailov, Kh. M., Osipov, O. A., Garnovskii, A. D., Kashireninov, O. E., Chikina, N. L. (Dokl. Akad. Nauk Azerb.SSR 21 [1965] 34/8).

[100] Gentile, P. S., Campiri, L. S., Carfagno, P. (J. Inorg. Nucl. Chem. 28 [1966] 1143/8).

[101] Molodkin, A. K., Ivanova, O. M., Kuchumova, A. N., Kozina, L. E. (Zh. Neorgan. Khim. 12 [1967] 1831/9; Russ. J. Inorg. Chem. 12 [1967] 963/7).

[102] Molodkin, A. K., Ivanova, O. M., Kozina, L. E. (Zh. Neorgan. Khim. 13 [1968] 2308/9; Russ. J. Inorg. Chem. 13 [1968] 1192/3).

[103] Petrov, K. I., Molodkin, A. K., Ivanova, O. M., Saralidze, O. D. (Zh. Neorgan. Khim. 14 [1969] 419/28; Russ. J. Inorg. Chem. 14 [1969] 215/9).

[104] Nefedov, V. I., Molodkin, A. K., Salyn, Ya. V., Ivanova, O. M., Parai-Koshits, M. A., Balakaeva, T. A., Belyakov, Z. V. (Zh. Neorgan. Khim. 19 [1974] 2628/31; Russ. J. Inorg. Chem. 10 [1974] 1435/7).

[105] Bagnall, K. W., Lopez, O. V. (J. Chem. Soc. Dalton Trans. 1975 1409/12).

[106] Bagnall, K. W., Lopez, O. V., Brown, D. (J. Inorg. Nucl. Chem. 38 [1976] 1997/8).

[107] Gritzner, G., Gutmann, V., Michalmayr, M. (Z. Anal. Chem. 224 [1967] 245/51).

[108] Bull, W. E., Madan, S. K., Willis, J. E. (Inorg. Chem. 2 [1963] 303/6).

[109] Bagnall, K. W., Brown, D., Jones, P. J., Robinson, P. S. (J. Chem. Soc. 1964 2531/4).

[110] Al-Daher, Abdul, G. M., Bagnall, K. W., Payne, G. F. (J. Less-Common Metals 115 [1986] 287/94).

[111] Temerk, Y. M., Ahmed, Z. A. (Bull. Soc. Chim. Belges 89 [1980] 15/24).

[112] Issa, I. M., Temerk, Y. M., Abu Zhuri, A. Z., El-Meligy, M. S., Kamal, M. M. (Monatsh. Chem. 111 [1980] 1151/8).

[113] Srivastava, A. K., Srivastava, M., Agarwal, R. K. (Ind. Eng. Chem. Prod. Res. Develop. 21 [1982] 135/7).

[114] Agarwal, R. K., Srivastava, A. K., Srivastava, T. N. (Proc. Natl. Acad. Sci. India A 51 [1981]; C.A. 98 [1983] No. 190594).

[115] Chacko, J. (Syn. React. Inorg. Metal-Org. Chem. 12 [1982] 361/71).

[116] Zhivopirtsev, V. P., Ryatosin, L. P. (Uch. Zap. Permsk. Gos. Univ. No. 207 [1970] 184/91; C.A. 76 [1972] No. 107411).

[117] Klimes, J., Knoechel, A., Rudolph, G. (Inorg. Nucl. Chem. Letters 13 [1977] 45/52).

[118] Jin, J. N., Xu, S. C., Liu, M. Z., Xu, D. Q., Zhou, M. L., Peng, Q. X. (He Huaxue Yu Fangshe Huaxue 3 [1981] 136/40).

[119] Costes, R. M., Folcher, G., Plurien, P., Rigny, P. (Inorg. Nucl. Chem. Letters 12 [1976] 13/21).

[120] Zhou, M. L., Jin, J. N., Liu, M. Z., Shi, M. X. (He Huaxue Yu Fangshe Huaxue 1 [1979] 103/4; C.A. 92 [1980] No. 120988).

[121] Wang, W., Chen, B., Wang, A., Yu., M., Liu, X. (He Huaxue Yu Fangshe Huaxue 4 [1982] 139/46; C.A. 97 [1982] No. 151583).

[122] Costes, R. M., Folcher, G., Plurien, P., Rigny, P. (Inorg. Nucl. Chem. Letters 12 [1976] 491/9).

[123] Tan, M., Fan, Y. (He Huaxue Yu Fangshe Huaxue 7 [1985] 165/70, 181).

[124] Rickard, C. E. F., Woollard, D. C. (Inorg. Nucl. Chem. Letters 14 [1978] 207/10).

[125] Agarwal, R. K., Srivastava, A. K., Srivastava, T. N. (Transition Metal Chem. [Weinheim] 5 [1980] 95/8).

[126] Srivastava, A. K., Agarwal, R. K. (Thermochim. Acta 56 [1982] 247/52).

[127] Bolle, J., Tomaszewski, J. (Fr. 1383992 [1963/65]; C.A. 62 [1965] 11455).

[128] Agarwal, R. K., Jain, P. C., Kapur, V., Sharma, S., Srivastava, A. K., (Transition Metal Chem. [Weinheim] 5 [1980] 237/9).

[129] Agarwal, R. K., Srivastava, A. K., Srivastava, T. N. (J. Inorg. Nucl. Chem. 42 [1980] 1366/8).

[130] Srivastava, A. K., Sharma, S., Agarwal, R. K. (Inorg. Chim. Acta 61 [1982] 235/9).

[131] Grunduz, N., Smith, B. C., Wassef, M. A., (Commun. Fac. Sci. Univ. Ankara 13 [1968] 693/4).

[132] Alvey, P., Bagnall, K. W., Brown, D. (J. Chem. Soc. Dalton Trans. 1973 2326/30).

[133] Molodkin, A. K., Ivanova, O. M., Belyakova, Z. V., Kolesnikova, L. E. (Zh. Neorgan Khim. 15 [1970] 3245/8; Russ. J. Inorg. Chem. 16 [1976] 1692/3).

[134] Abraham, J., Corsini, A. (Anal. Chem. 42 [1970] 1528/31).

[135] Cotton, F. A., Francis, R. (J. Am. Chem. Soc. 82 [1960] 2986/91).

[136] Petrov, K. I., Ivanova, O. M., Molodkin, A. K. (Zh. Neorgan. Khim. 17 [1972] 1613/5; Russ. J. Inorg. Chem. 17 [1972] 834/6).

[137] Rozen, A. M., Murinov, Yu. I., Nikitin, Yu. E., Teterin, E. G., Kaminskaya, L. K., Mazalov, L. N., Gal'tseva, E. A. (Radiokhimiya 15 [1973] 123/5; Soviet Radiochem. 15 [1973] 123/5).

[138] Nikolaev, A. V., Torgov, V. G., Andrievskii, V. M., Gal'tsova, E. A., Gil'bert, E. H., Kotlyarovskii, I. L., Mazalov, L. M., Mikhailov, V. A., Cheremisina, I. M. (Zh. Neorgan. Khim. 15 [1970] 1336/42; Russ. J. Inorg. Chem. 15 [1970] 685/9).

[139] Alvey, P., Bagnall, K. W., Velasquez, O. L., Brown, D. (J. Chem. Soc. Dalton Trans. 1975 1277/80).

[140] Mirra, B. B., Moharty, S. R., Murti, M. V. V. S., Raychandhuri, S. (Inorg. Chim. Acta 28 [1978] 275/81).

[141] Nikolaev, A. V., Khudorozhko, G. F., Mazalov, L. N., Yumatov, V. D., Torgov, V. G., Drozdova, M. K. (Dokl. Akad. Nauk SSSR 233 [1977] 133/6; Dokl. Chem. Proc. Acad. Sci. USSR 232/237 [1977] 122/4).

[142] Gadia, M. K., Rai, R. S. (Indian J. Chem. 11 [1973] 392/3).

[143] Savant, V. V., Patel, C. C. (J. Less-Common Metals 24 [1971] 459/65).

[144] Ramalingam, S. K. (Proc. 1st Chem. Symp., Chandigarrh, India, 1969 [1970], Vol. 2, pp. 308/11).

[145] Srivastava, A. K., Agarwal, R. K. (Thermochim. Acta 56 [1982] 247/52).

[146] Bagnall, K. W., Wakerley, M. W. (J. Chem. Soc. Dalton Trans. 1974 889/95).

[147] Alcock, N. W., Esperas, S., Bagnall, K. W., Wang, H. Y. (J. Chem. Soc. Dalton Trans. **1978** 638/46).
[148] Vdovenko, V. M., Kovalaeva, T. V., Kanevskaya, N. A., Stantsevich, V. P., Suglobov, D. N. (Radiokhimiya **14** [1972] 134/6; Soviet Radiochem. **14** [1972] 135/7).
[149] English, R. P., du Preez, J. G. H., Nassimbeni, L. R. (S. African J. Chem. **32** [1979] 119/25).
[150] du Preez, J. G. H., van Vuuren, C. P. J., McGill, W. J. (J. Coord. Chem. **5** [1975/76] 231/5).

[151] du Preez, J. G. H., van Vuuren, C. P. J. (J. Chem. Soc. Dalton Trans. **1975** 1548/52).
[152] Mazhar-Ul-Haque, Laughlin, C. N., Hart, F. A., van Nice, R. (Inorg. Chem. **10** [1971] 115/22).
[153] Parker, J. R., Banks, C. V. (IS-982 [1964] 1/185; C.A. **62** [1965] 9857).
[154] Parker, J. R., Banks, C. V. (J. Inorg. Nucl. Chem. **27** [1965] 583/7).

4.6.5 Solutions of $Th(NO_3)_4$ and $Th(NO_3)_4$ Hydrates

Vapour pressures of thorium nitrate aqueous solutions have been reported by Robinson, Levinson [1].

Information on vibrational spectra of thorium nitrate recorded in aqueous media is available in [2, 4, 7, 10 to 14] and details of NMR investigations of similar solutions will be found, for example, in [15 to 17]. An NMR study of $Th(NO_3)_4 \cdot 4H_2O$ in acetone-acetonitrile mixture has also been reported [18].

Infrared spectral data have also been reported for solutions of thorium tetranitrate hydrates in solvents such as acetone [5 to 7], diethyl ether [3, 7], dioxane, acetonitrile and dimethylformamide [7], methylisobutylketone, dibutyl "carbitol" and tributylphosphate [8] and tri-n-octylphosphine oxide [9]. The results indicate the presence of co-ordinated nitrate groups.

References for 4.6.5:

[1] Robinson, R. A., Levinson, B. J. (Trans. Proc. Roy. Soc. New Zealand **76** [1946] 295/9).
[2] Mathien, J. P., Lounsbury, M. (Discussions Faraday Soc. No. 9 [1950] 196/207).
[3] Ryskin, Ya. I., Shvedov, V. P., Solov'eva, A. A. (Zh. Neorgan. Khim. **4** [1959] 2268/76; Russ. J. Inorg. Chem. **4** [1959] 1033/7).
[4] Ferraro, J. R. (J. Inorg. Nucl. Chem. **10** [1959] 319/22).
[5] Norwitz, G., Chasan, D. E. (J. Inorg. Nucl. Chem. **31** [1969] 2267/70).
[6] Shevchenko, L. L., Shurkal, T. M. (Ukr. Khim. Zh. **36** [1970] 199/203; Soviet Progr. Chem. **36** No. 2 [1970] 90/4; C.A. **73** [1970] No. 8993).
[7] Vdovenko, V. M., Statsevich, V. P., Suglobov, D. N. (Radiokhimiya **14** [1972] 366/70; Soviet Radiochem. **14** [1972] 377/80).
[8] Katzin, L. I. (J. Inorg. Nucl. Chem. **24** [1962] 245/56).
[9] Verstegen, J. M. P. J. (J. Inorg. Nucl. Chem. **26** [1964] 25/35).
[10] Hester, R. E., Plane, R. A. (J. Chem. Phys. **40** [1964] 411/4).

[11] Hester, R. E., Plane, R. A. (Inorg. Chem. **3** [1964] 769/70).
[12] Volod'ko, L. V., Lieh Than Huoah (Zh. Prikl. Spektrosk. **9** [1968] 644/9; J. Appl. Spectrosc. [USSR] **9** [1968] 1100/4).
[13] Oliver, B. G., Davis, A. R. (J. Inorg. Nucl. Chem. **34** [1972] 2851/60).

[14] Alves-Marques, M., Oksengorn, B., Vodas, B. (Advan. Raman Spectrosc. **1** [1972] 585/92).
[15] Axtmann, R. C. (J. Chem. Phys. **30** [1959] 340/1).
[16] Berger, C., Emans, H. H., Pohl, D. (Z. Physik. Chem. [Leipzig] **256** [1975] 421/9).
[17] Chew, F. K., Healy, M. A., Khalil, I. M., Logan, N. (J. Chem. Soc. Dalton Trans. **1975** 1315/8).
[18] Buslaev, Yu. A., Petrosyants, S. P., Buslaeva, N. M., Tarasov, V. P. (Dokl. Akad. Nauk SSSR **187** [1969] 120/3).

4.6.6 Thorium Oxide Nitrate, ThO(NO₃)₂, and Oxide Nitrate Hydrates

Preparation. Thermal decomposition of $Th(NO_3)_4 \cdot 5H_2O$ in air (2°C/min) has yielded a product of composition close to $ThO(NO_3)_2$ at 194°C [2, 3]. According to [1] the oxide nitrate may be prepared by thermal decomposition of $(NO_2)_2Th(NO_3)_6$ at 190°C in vacuum; at a lower temperature (150°C) the product is $Th(NO_3)_4$, cf. p. 70.

The earlier work on $ThO(NO_3)_2 \cdot 0.5H_2O$ is covered in "Thorium" 1955, p. 251. Subsequently Vdovenko et al. [4] reported the formation of a phase of composition $ThO(NO_3)_{2.2} \cdot H_2O$ on thermal decomposition of $Th(NO_3)_4 \cdot 5H_2O$ in a vacuum (10^{-5} Torr) at 140°C over a period of 16 h.

Claudel [2] prepared the hydrated phase $ThO(NO_3)_2 \cdot 2H_2O$ by dissolving hydrated thorium hydroxide in the calculated quantity of 2M nitric acid ($Th:NO_3 = 1:2$) and evaporating the resulting solution to dryness. No evidence for the presence of a Th–O bond was provided and it is possible that the product was a hydroxide nitrate hydrate rather than an oxide nitrate hydrate, or even a mixture of different phases. The water of hydration is reported to be lost between ambient temperature and 120°C on heating at 2°C/min.

Properties. Nitrate vibrations have been recorded in the IR spectrum of $ThO(NO_3)_{2.2} \cdot H_2O$ at 1610, 1230, 980, 794, and 744 cm^{-1}. On the basis of the large difference between the first two of these bands it is suggested that bridging nitrates are present [4].

A complex with trimethylphosphate (L), $ThO(NO_3)_2 \cdot 2L$, is reported to form when $Th(NO_3)_4 \cdot 3L$ is stored over phosphorus pentoxide [5]. A further oxide nitrate complex of composition $[\{ThL'(NO_3)H_2O\}_2O](NO_3)_4 \cdot H_2O$ (L' = 2,9-diformyl-1,10-phenanthrolinedisemicarbazone) has recently been isolated from ethanol solution and characterized crystallographically [6].

References for 4.6.6:

[1] Schmeisser, M., Koehler, G. (Angew. Chem. **77** [1965] 456).
[2] Claudel, B. (Diss. Univ. Lyon 1962, pp. 1/95).
[3] Bussière, P., Claudel, B., Renouf, J. P., Trambouze, Y., Prettre, M. (J. Chim. Phys. **58** [1961] 668/74).
[4] Vdovenko, V. M., Statsevich, V. P., Suglobov, D. H. (Radiokhimiya **8** [1966] 211/7; Soviet Radiochem. **8** [1966] 194/9).
[5] Legin, E. K., Malikina, G. A., Shpunt, L. B., Shcherbakova, L. L. (Radiokhimiya **22** [1980] 300/3).
[6] Aghabozorg, H., Palenik, R. C., Palenik, G. J. (Inorg. Chim. Acta **76** [1983] L259/L260).

4.6.7 Thorium Peroxide Nitrate Hydrates

Thorium peroxide nitrate phases of indefinite composition were reported by Hammaker, Koch [1]. Subsequently, Johnson et al. [2] showed that mixing hot dilute nitric acid solutions of $Th(NO_3)_4 \cdot 5H_2O$ and hydrogen peroxide gave a solid which, after being washed with hot water and acetone, and vacuum dried at room temperature over $Mg(ClO_4)_2$, analysed reproducibly as $Th_6(O_2)_{10}(NO_3)_4 \cdot 10H_2O$. This white solid is very stable and ca. 30 min are required for completion of ceric sulfate titrations in hot $1M$ H_2SO_4.

The effect of temperature on peroxide decomposition in a $2M$ $Th(NO_3)_4$ solution in $2.2M$ HNO_3 is shown in **Fig. 27**; the activation energy is 18 ± 2 kcal/mol [2].

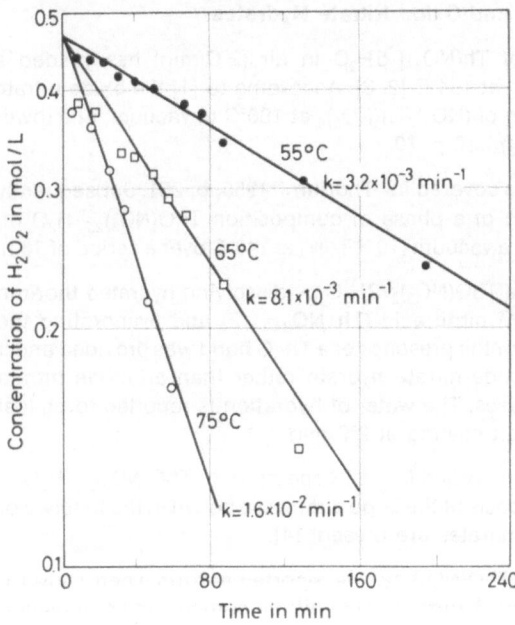

Fig. 27

Effect of temperature on peroxide decomposition in a solution of $Th_6(O_2)_{10}(NO_3)_4 \cdot 10H_2O$, $2M$ in $Th(NO_3)_4$ and $2.2M$ in HNO_3 [2].

The peroxide nitrate is stable for at least 2 months on storage in a desiccator [2]. Thermal decomposition yields ThO_2 at elevated temperature [1].

References for 4.6.7:

[1] Hammaker, J. W., Koch, C. W. (TID-5223 [1952] 318/38; N.S.A. 11 [1957] No. 11493).
[2] Johnson, G. L., Kelly, M. J., Cuneo, D. R. (J. Inorg. Nucl. Chem. 27 [1965] 1787/91).

4.6.8 Thorium Hydroxide Nitrate Hydrates

Preparation

Although there has been considerable interest in the soluble hydrolysed species in aqueous thorium nitrate solutions (see, for example, [1 to 6]) few solid hydroxide nitrate compounds have been isolated (Table 18, p. 70) and of these only one has been fully characterized. Thus

References for 4.6.8 see p. 106

the hydrated compound $Th_2(OH)_2(NO_3)_6 \cdot 8H_2O$ crystallises from weakly hydrolysed thorium nitrate solutions [2, 10] whilst a phase of composition $Th(OH)_2(NO_3)_2 \cdot xH_2O$ is formed on further hydrolysis [6]. According to Deptula [2] x = 3 for the latter compound. These products were obtained by room temperature evaporation of solutions containing, respectively, OH:Th ratios of 0.5 and 1.0 [6] and by dissolution of 3:1 and 2:1 mole ratios of thorium hydroxide in 1M thorium nitrate solution followed by evaporation in a stream of nitrogen [2]. The more extensively hydrolysed species $Th(OH)_3(NO_3) \cdot xH_2O$ (x = 2 and 1.5) and $ThO_x(OH)_y(NO_3)_{4-2x-y}$ have been isolated by addition of different amounts of concentrated aqueous ammonia (25%) to a Th^{IV} solution in 0.4M HNO_3 (100 g Th/L) [7] (see also [8]). Deptula [2] reports the formation of phases containing ca. 1 and 0.5 nitrate molecules per thorium atom but does not give definite formulae for them whilst according to [9] a phase of composition $Th(OH)_{3.7}(NO_3)_{0.3}$ is formed on addition of aqueous sodium hydroxide to aqueous thorium tetranitrate solution.

Physical Properties

The available X-ray crystallographic data are provided in Table 39, p. 102. Johansson [10] has reported the structure of $Th_2(OH)_2(NO_3)_6 \cdot 8H_2O$, which contains discrete dinuclear units, $Th_2(OH)_2(NO_3)_6(H_2O)_6$, with the remaining water molecules being involved in hydrogen bonding.

A projection of the structure along the b axis is shown in **Fig. 28** and the atomic arrangement is illustrated in **Fig. 29**. Bond distances are provided in Table 40, p. 103, from which it will be noted that the Th–Th distance is 398.8 pm. The Th atoms are joined by two hydroxo-bridges and the co-ordination number of 11 is completed by 6 oxygen atoms from 3 bidentate nitrate groups and 3 from bonded water molecules. The resulting irregular co-ordination polyhedron can be described as a somewhat distorted dodecahedron if each nitrate group is considered as a single unit [10].

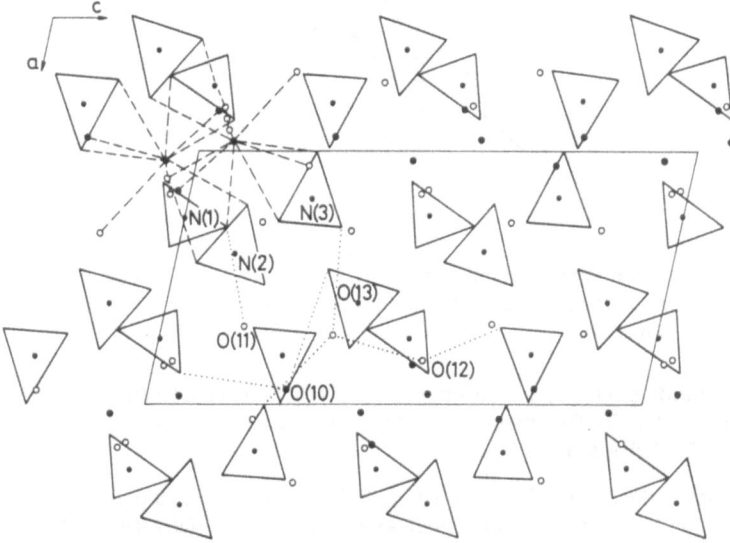

Fig. 28

A projection of the structure of $Th_2(OH)_2(NO_3)_6 \cdot 8H_2O$ along the b axis. All Th–O contacts within one of the dinuclear complexes are marked by dashed lines. The dotted lines indicate some of the hydrogen bonds [10].

            References for 4.6.8 see p. 106

Table 39
Crystallographic Data for Thorium Hydroxide Nitrate Hydrates.

compound	colour	symmetry; space group	lattice parameters in pm (or °)				Ref.
			a	b	c	β	
$Th_2(OH)_2(NO_3)_6 \cdot 8H_2O$	white	monoclinic; $P2_1/c$ (No. 14)	677.2	1169.3	1376.9	102.63°	[6, 10]
$Th(OH)_2(NO_3)_2 \cdot xH_2O^{*)}$	white	monoclinic; $C2/c$ (No. 15)	1425	895	611	98.0°	[6]

*) x = 3 according to Deptula [2].

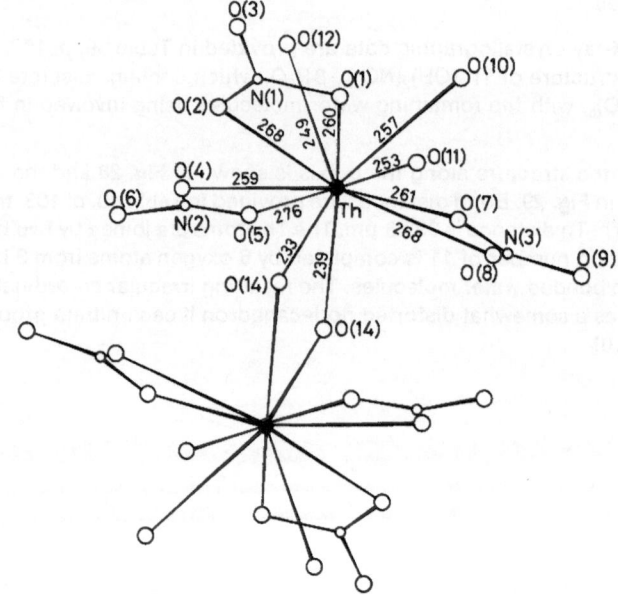

Fig. 29
The atomic arrangement in the $Th_2(OH)_2(NO_3)_6(H_2O)_6$ complex as
viewed down [100], with [010] vertical (interactomic distances
in pm) [10].

Structural studies have shown [6] that the Th atoms in $Th(OH)_2(NO_3)_2 \cdot xH_2O$ form infinite
zig-zag chains with a Th–Th distance of 400 pm; the structure refinement was not as success-
ful as that for $Th_2(OH)_2(NO_3)_6 \cdot 8H_2O$.

The calculated density for $Th_2(OH)_2(NO_3)_6 \cdot 8H_2O$ (X-ray) is 3.209 g/cm³; the experimentally
determined value is 3.20 g/cm³ [10]. Densities reported for $Th(OH)_3(NO_3) \cdot 2H_2O$ and $Th(OH)_3$-
$(NO_3) \cdot 1.5H_2O$ are 3.45 and 3.44 g/cm³, respectively [7].

Infrared assignments for the compounds $Th_2(OH)_2(NO_3)_6 \cdot 8H_2O$ and $Th(OH)_2(NO_3)_2 \cdot 3H_2O$
are given in Table 41 [2]. Less detailed information on phases of composition $Th(OH)_3(NO_3)$
$\cdot nH_2O$ is tabulated by Shmidt et al. [7].

References for 4.6.8 see p. 106

Table 40

Distances (in pm) within the $Th_2(OH)_2(NO_3)_6(H_2O)_6$ Group [10].

Th–Th 398.8 pm

Distances within the first coordination sphere of a Th atom:

nitrate oxygens			water oxygens		
Th–O(1)	260		Th–O(10)	257	
Th–O(2)	268		Th–O(11)	253	
Th–O(4)	259		Th–O(12)	249	
Th–O(5)	276		hydroxo oxygens		
Th–O(7)	261		Th–O(14)	239	
Th–O(8)	268		Th–O(14)	233	
			O(14)–O(14)	252	

Other distances involving thorium atoms:

Th–N(1)	308	Th–O(9)	431	Th–O(7)	435
Th–N(2)	313	Th–O(1)	540	Th–O(8)	456
Th–N(3)	309	Th–O(2)	505	Th–O(10)	626
Th–O(3)	425	Th–O(4)	415	Th–O(11)	565
Th–O(6)	432	Th–O(5)	476	Th–O(12)	620

Distances within the nitrate groups:

N(1)–O(1)	128	N(2)–O(4)	126	N(3)–O(7)	130
N(1)–O(2)	123	N(2)–O(5)	126	N(3)–O(8)	127
N(1)–O(3)	118	N(2)–O(6)	120	N(3)–O(9)	122
O(1)–O(2)	212	O(4)–O(5)	212	O(7)–O(8)	218
O(1)–O(3)	213	O(4)–O(6)	216	O(7)–O(9)	217
O(2)–O(3)	212	O(5)–O(6)	215	O(8)–O(9)	221

Possible hydrogen bond distances:

O(10)–O(2)	293	O(12)–O(9)	277	O(13)–O(9)	296
O(10)–O(6)	284	O(12)–O(13)	277	O(13)–O(11)	283
O(11)–O(1)	290	O(13)–O(8)	281	O(13)–O(12)	277
O(11)–O(13)	283				

Table 41

Infrared Data for Thorium Hydroxide Nitrate Hydrates (in cm^{-1}) [2].
s = strong; m = medium; w = weak; vw = very weak; sh = shoulder;
b = broad.

assignment	$Th_2(OH)_2(NO_3)_6 \cdot 8 H_2O$	$Th(OH)_2(NO_3)_2 \cdot 3 H_2O$
hydrogen bonded νH_2O and OH st.	3580 sh	3510 s, b
	3510 s	3340 s, b
	3480 s	3190 b

References for 4.6.8 see p. 106

Table 41 (continued)

assignment	$Th_2(OH)_2(NO_3)_6 \cdot 8\,H_2O$	$Th(OH)_2(NO_3)_2 \cdot 3\,H_2O$
	3380 s	—
	3360 s	—
	3220 b	—
$\nu_1 + \nu_2$	2535 w	2520 w
	2510 w	2500 w
$\nu_2 + \nu_4$	2390 w	2370 w
	2295 w	2215 w
	2250 vw, sh	—
$2\nu_2$	2075 vw	2045 vw
	2050 vw	—
$\nu_2 + \nu_3$	1810 sh, vw	1780 vw
	1785 vw	—
	1780 vw	—
$\nu_2 + \nu_5$	1755 vw	1760 vw
	1740 vw	—
δH_2O	1640 s	1643 s
	1625 s	1632 s
ν_1	1560 sh	1540 sh
	1535 sh	1525 s
	1528 s	1515 s
	1500 sh	1500 sh
ν_4	1367 sh	1305 s
	1347 s	—
	1314 s	—
	1288 s	—
ν_2	1058 sh	1055 sh, w
	1046 sh	1040 s
	1037 sh	—
	1032 s	—
δOH	843 m	840 m
ν_6	809 m	813 m
ν_3	758 sh	743 m
	751 m	—

References for 4.6.8 see p. 106

Chemical Properties

Thermal decomposition of $Th(OH)_3(NO_3) \cdot xH_2O$ (x = 1.5 and 2) in air (heating rate 5°C/min^{-1}) is accompanied by three endothermal effects and one isothermal effect. It is suggested that the decomposition proceeds via the formation of $ThO(OH)(NO_3)$ at 225°C [7]. The ultimate product is ThO_2 [7, 11].

Solubility data for the system $Th(NO_3)_4$–HNO_3–NH_4OH–H_2O, obtained by addition of different amounts of 25% aqueous ammonia to Th^{IV} in 0.4 M HNO_3 (100 g/L) at 19 to 21°C are listed in Table 42 and the compositions of the precipitated phases are shown in Table 43. Those for the first and third experiment correspond closely to $Th(OH)_3(NO_3) \cdot 1.5H_2O$ and $Th(OH)_3(NO_3) \cdot 2H_2O$, respectively [7].

Table 42

Solubility of Thorium Compounds in the System $Th(NO_3)_4$–HNO_3– NH_4OH–H_2O [7].

volume of 25% ammonia solution added as % of total volume	pH of mother liquor	thorium concentration in mother liquor in g/L
15.1	3.80	50.0
16.0	3.90	26.0
16.2	4.10	18.2
16.6	6.05	0.460
17.0	7.10	0.007
20.0	8.85	0.150
30.0	9.50	0.006
46.0	10.10	0.004
60.0	10.25	0.002
77.0	10.38	0.001

Although not prepared directly from a hydroxide nitrate the complex $Th(OH)_3(NO_3)(NO)$ $\cdot H_2O$ is reported to form when nitric oxide is bubbled through a suspension of $Th(OH)_4 \cdot xH_2O$ in ethanol [12]. Infrared data are listed in Table 44, p. 107.

Table 43

Composition and Density of Solid Phases in the System $Th(NO_3)_4$–HNO_3–NH_4OH–H_2O [7].

pH	density in g/cm³	found %					
		Th	NO_3^-	OH^-	O^{2-}	NH_4^+	H_2O
3.80	3.44	62.5	16.4	13.7	—	—	7.4
3.90	3.45	61.8	16.0	13.3	—	—	8.9
4.10	3.45	60.5	17.0	12.8	—	—	9.7

                        References for 4.6.8 see p. 106

Table 43 (continued)

pH	density in g/cm³	found %					
		Th	NO$_3^-$	OH$^-$	O^{2-}	NH$_4^+$	H$_2$O
6.05	3.21	59.0	15.6	6.0	2.8	0.5	16.1
7.10	2.45	40.5	14.5	5.3	2.1	1.2	36.4
8.85	2.39	42.7	12.9	5.4	2.5	1.5	35.0
9.50	3.33	58.4	7.5	9.5	2.9	0.2	21.5
10.10	3.48	65.2	7.5	9.7	3.0	0.2	14.4
10.25	3.64	67.4	7.1	8.9	3.4	0.2	13.0
10.38	3.81	68.5	4.7	7.5	4.1	—	15.2

References for 4.6.8:

[1] Johansson, G. (Acta Chem. Scand. **22** [1968] 399/409).
[2] Deptula, A. (Nukleonika **17** [1972] 47/58).
[3] Magini, M., Cabrini, A., Scibona, G., Johansson, G., Sandstrom, M. (Acta Chem. Scand. A **30** [1976] 437/47).
[4] Magini, M., Cabrini, A., Di Bartolomeo, A. (RT-CHI-75-2 [1975] 1/60; C.A. **84** [1976] No. 35711).
[5] Magini, M., Cabrini, A., Di Bartolomeo, A. (CNEN-RT-CHI-75-2 [1975] 1/43; C.A. **86** [1977] No. 10942).
[6] Johansson, G. (Svensk Kem. Tidskr. **78** [1966] 486/9).
[7] Shmidt, V. S., Andryushin, V. G., Slepchenko, I. G., Teterin, E. G., Shesterikov, N. N., Chuklinov, R. N. (Radiokhimiya **21** [1979] 642/9; Soviet Radiochem. **21** [1979] 558/64).
[8] Mzareulishvili, N. V. (Soobshch. Akad. Nauk Gruz.SSR **26** [1961] 653/8; C.A. **56** [1962] 3101).
[9] Singh, R. P., Banerjee, N. R. (J. Indian Chem. Soc. **39** [1962] 255/9).
[10] Johansson, G. (Acta Chem. Scand. **22** [1968] 389/98).

[11] Shmidt, V. S., Andryushin, V. G. (Radiokhimiya **24** [1982] 601/6; Soviet Radiochem. **24** [1982] 498/503).
[12] Molodkin, A. K., Ivanova, O. M., Goeva, L. V., Balakaeva, T. A., Belyakova, Z. V., Chaplya, T. A. (Zh. Neorgan. Khim. **25** [1980] 1907/10; Russ. J. Inorg. Chem. **25** [1980] 1057/9).

4.6.9   Thorium Hyponitrite-Nitrates

Nitrosylation of an aqueous solution of Th(NO$_3$)$_4$·5H$_2$O is reported to yield Th(NO$_3$)$_2$(N$_2$O$_2$)(NO)·2H$_2$O whilst similar reactions in alcohol yield either Th(NO$_3$)$_2$(N$_2$O$_2$)·4H$_2$O (after 18 h) or Th(NO$_3$)$_2$(N$_2$O$_2$)(NO)$_2$·3H$_2$O (after 25 to 28 h). The basic compound Th(OH)(NO$_3$)(N$_2$O$_2$)·3H$_2$O is apparently formed on nitrosylation of Th(OH)$_4$·xH$_2$O suspended in water; the same reaction in ethanol yields Th(OH)$_3$(NO$_3$)(NO)·H$_2$O. Infrared assignments are listed in Table 44.

It would be interesting to have additional information on these compounds and on those containing the chloride ion (e.g., ThCl$_3$(NO$_3$)(NO)$_2$·2H$_2$O).

Table 44

Infrared Assignments (in cm^{-1}) for Thorium Hyponitrite-Nitrate Hydrates.

| compound | $\nu(H_2O)$ | $\delta(H_2O)$ | NO$_3^-$ | | | | | N$_2$O$_2^{2-}$ | $\nu(NO)$ |
			$\nu_4(B_2)$	$\nu_1(A_1)$	$\nu_2(A_1)$	$\nu_6(B_1)$	$\nu_3(A_1)$	$\nu(NO)$	
Th(OH)(NO$_3$)-(N$_2$O$_2$)·3H$_2$O	3600 to 3100	1645	1500	1315	1042	820	755	1170	—
Th(OH)$_3$(NO$_3$)-(NO)·H$_2$O	3600 to 3100	1640	1520	1310	1042	820	755	—	1655 1670
Th(NO$_3$)$_2$(N$_2$O$_2$)-(NO)·2H$_2$O	3600 to 3100	1640	1520	1290	1035	810	750	992 960	1620
Th(NO$_3$)$_2$(N$_2$O$_2$)-(NO)$_2$·3H$_2$O	3600 to 3100	1640	1520	1320	1045	810	750	1170	1620
Th(NO$_3$)$_2$(N$_2$O$_2$)·4H$_2$O	3600 to 3100	1640	1520	1340 1320	1045	812	755	1030	—

Reference for 4.6.9:

Molodkin, A. K., Ivanova, O. M., Goeva, L. V., Balakaeva, T. A., Belyakova, Z. V., Chaplya, T. A. (Zh. Neorgan. Khim. **25** [1980] 1907/10; Russ. J. Inorg. Chem. **25** [1980] 1057/9).

4.7 Thorium Double Nitrates or Nitrato Complexes

David Brown
Chemistry Division, Atomic Energy Establishment
Harwell, England

The thorium double nitrates or nitrato complexes are listed in Table 45, p. 108. Pre-1950 publications are dealt with in "Thorium" 1955, pp. 323, 329/30, 337, 343/6 and 349/51 and will only be mentioned where appropriate in the current article.

4.7.1 Preparation

Pentanitrato Complexes. The pentakis(nitrato) complexes NaTh(NO$_3$)$_5$·8.5H$_2$O and KTh(NO$_3$)$_5$·6H$_2$O have been obtained by crystallisation from aqueous nitric acid in a desiccator over concentrated sulfuric acid; the products were dried between sheets of filter paper [1]. Stable lower hydrates were not characterized in thermogravimetric studies. The analogous complexes containing alkylammonium cations RTh(NO$_3$)$_5$·xH$_2$O (R=CH$_3$NH$_3^+$, x=7; R= (CH$_3$)$_2$NH$_2^+$, x=8; R=C$_2$H$_5$NH$_3^+$, x=5 and R=(C$_2$H$_5$)$_2$NH$_2^+$, x=11) crystallise readily when aqueous solutions containing thorium tetranitrate and the appropriate alkylammonium nitrate are kept in a vacuum desiccator over potassium hydroxide [2]. Evaporation of an acetone solution of P(C$_6$H$_5$)$_4$NO$_3$ and Th(NO$_3$)$_4$·2.33(CH$_3$)$_3$PO has yielded the complex [P(C$_6$H$_5$)$_4$]- [Th(NO$_3$)$_5$((CH$_3$)$_3$PO)$_2$]; the product was recrystallised from acetone-methanol mixture and dried in a vacuum [18].

Table 45
Thorium(IV) Nitrato Complexes[a].

pentakis (nitrato) complexes[a]	hexakis (nitrato) complexes[b,c]	heptakis (nitrato) complexes	miscellaneous
$NaTh(NO_3)_5 \cdot xH_2O$ (x = 9, 8.5, 7, 2)	$(M^I)_2Th(NO_3)_6$ (M^I = K, Rb, Cs)	$K_3Th(NO_3)_7$	$K_3H_4[Th(NO_3)_{11}]$[d]
	$M^{II}Th(NO_3)_6$ (M^{II} = Mg, Zn, Mn, Co, Ni)		
$KTh(NO_3)_5 \cdot xH_2O$ (x = 9, 7, 6, 1.5)	$(R^I)_2Th(NO_3)_6$ (R^I = NH_4^+, $(CH_3)NH_3^+$, $(CH_3)_2NH_2^+$, $(C_2H_5)_2NH_2^+$, $(C_2H_5)_4N^+$, $(CH_3)_2(C_6H_5CH_2)NH^+$, $(CH_3)_2(C_6H_5CH_2)_2N^+$, $(CH_3)_3(C_6H_5CH_2)N^+$, $(CH_3)_2(C_2H_5)(C_6H_5CH_2)N^+$, $(CH_3)_2(C_3H_7)(C_6H_5CH_2)N^+$, $(CH_3)_2(C_4H_9)(C_6H_5CH_2)N^+$, $C_5H_5NH^+$, $CH_3 \cdot C_5H_4NH^+$, $C_2H_5 \cdot C_5H_4NH^+$, $C_3H_7 \cdot C_5H_4NH^+$, $C_4H_9 \cdot C_5H_4NH^+$)		
$NH_4Th(NO_3)_5 \cdot xH_2O$ (x = 8, 5, 1)			
$[(CH_3)NH_3]Th(NO_3)_5 \cdot 7H_2O$			
$[(CH_3)_2NH_2]Th(NO_3)_5 \cdot 8H_2O$	$(NO)_2Th(NO_3)_6$		
$[(C_2H_5)NH_3]Th(NO_3)_5 \cdot 5H_2O$	$(NO_2)_2Th(NO_3)_6$		
$[(C_2H_5)_2NH_2]Th(NO_3)_5 \cdot 11H_2O$	$[Th(NO_3)_3(tmpo)_4]_2[Th(NO_3)_6]$		
$[P(C_6H_5)_4][Th(NO_3)_5 \cdot (tmpo)_2]$	$[Th(NO_3)_3(hmpa)_4]_2[Th(NO_3)_6]$		
	$[(bpyH)_3NO_3][Th(NO_3)_6]$		
	$[(phenH)_3NO_3][Th(NO_3)_6]$		

a) The existence of complexes such as $NaTh_2(NO_3)_9 \cdot xH_2O$ (x = 20, 8) and $(NH_4)_4Th_5(NO_3)_{24} \cdot xH_2O$ (x = 25, 11, 9) (see "Thorium" 1955, pp. 323 and 337) has not been confirmed. — b) tmpo = trimethylphosphine oxide. — c) hmpa = hexamethylphosphoramide; bpy = 2,2'-bipyridine; phen = 1,10-phenanthroline. — d) It is possible that the compound reported as $K_3H_3[Th(NO_3)_{10}] \cdot 4H_2O$ or $K_3(H_3O)_3[Th(NO_3)_{10}] \cdot H_2O$ (see "Thorium" 1955, p. 330 and this article, p. 109) is identical with this phase.

References for 4.7 see p. 123

Hexanitrato Complexes. Alkali metal and ammonium hexanitrato complexes, $M'_2Th(NO_3)_6$ ($M' = Rb$, Cs, NH_4), crystallise from nitric acid solutions containing the appropriate cation and thorium nitrate [1, 3 to 7, 13]. The caesium salt is also obtained when $(NO_2)_2Th(NO_3)_6$ is reacted with $CsNO_3$. The nitronium complex, $(NO_2)_2Th(NO_3)_6$, is conveniently obtained by addition of chlorine nitrate to thorium tetrachloride [8] or by the addition of dinitrogen pentoxide to thorium tetranitrate tetrahydrate in anhydrous nitric acid [23]. Treatment with dinitrogen tetroxide results in the formation of $(NO)_2Th(NO_3)_6$ [8].

Anhydrous hexanitrato complexes containing substituted ammonium and pyridinium cations are readily obtained by crystallisation from aqueous nitric acid solutions of the appropriate nitrates; viz. $R_2Th(NO_3)_6$ (R = dimethylammonium, diethylammonium [2]; dimethylbenzyl-ammonium [10, 12]; dimethyldibenzylammonium [13]; trimethylbenzylammonium [10, 12, 13]; ethyldimethylbenzylammonium, propyldimethylbenzylammonium, butyldimethylbenzylammonium [12]; tetramethylammonium [9]; pyridinium, methylpyridinium, ethylpyridinium, propylpyridinium, butylpyridinium [11]); see also [14, 15]. The formation of $R_2Th(NO_3)_6$ (R = tri-n-octylammonium) in benzene solution is mentioned in [22].

Addition of methanolic 2,2'-bipyridine (= bpy) to a solution of thorium in nitric acid-tributylphosphate mixture obtained by electrochemical oxidation of thorium metal yields the unusual complex $[(bpyH)_3NO_3][Th(NO_3)_6]$ when the solution volume is reduced in a vacuum [24]. Similar reactions involving 1,10-phenanthroline (= phen) and tetraethylammonium nitrate yield $[(phenH)_3NO_3][Th(NO_3)_6]$ and $[N(C_2H_5)_4]_2[Th(NO_3)_6]$, respectively [24].

The compound $Th(NO_3)_4 \cdot 2.67(CH_3)_3PO$, which has been shown by a full structure determination to be $[Th(NO_3)_3(C_3H_9PO)_4]_2[Th(NO_3)_6]$ (see p. 112), is obtained by evaporation of an ethanol solution containing trimethylphosphine oxide (4.5 mmol) and hydrated thorium tetranitrate (1.5 mmol) followed by recrystallisation of the initial product from a 1:2 methylcyanide-methanol mixture [18]. The hexamethylphosphoramide (= hmpa) analogue $[Th(NO_3)_3(hmpa)_4]_2[Th(NO_3)_6]$ forms on addition of 2-methylbutane to an acetone solution of $Th(NO_3)_4 \cdot 5H_2O$ containing hexamethylphosporamide [19].

Hexanitrato complexes of the type $M''Th(NO_3)_6 \cdot 8H_2O$ form on mixing the appropriate divalent metal nitrate and thorium tetranitrate in aqueous nitric acid ($M'' = Mg$, Zn, Co, Ni [3]; $M'' = Mg$, Mn, Co, Ni, Zn [16]; $M'' = Mg$, Mn, Co, Ni [1]).

Other Nitrato Complexes. The early work on a phase of composition $3KNO_3–Th(NO_3)_4 \cdot 3HNO_3 \cdot 4H_2O$ is referred to in "Thorium" 1955, p. 330. Subsequently, Molodkin et al. [1] reported the preparation of $K_3(H_3O)_3[Th(NO_3)_{10}] \cdot H_2O$ which was later stated to be rhombohedral (pseudo cubic) with a = 1115.4 pm [25]. According to Douglass [17] the salt $K_3H_4[Th(NO_3)_{11}]$ is obtained from 16 M HNO_3 containing potassium and thorium nitrates and this is also stated to be rhombohedral with a = 1111(1) pm, $\alpha = 90.5(1)°$. Neither the analytical results for the former preparation nor the Th:K:NO$_3$ ratios given by Douglass [17] are convincing. Additional work will be of interest to resolve this situation.

The heptakis (nitrato) complex $K_3Th(NO_3)_7$ is apparently formed when $K_3(H_3O)_3[Th(NO_3)_{10}] \cdot H_2O$ is heated to constant weight at 150°C; infrared studies indicate the absence of ionic nitrate in the complex [1].

Information on nitrate glasses containing thorium and various other cations is available in [21].

References for 4.7 see p. 123

4.7.2 Physical Properties

Available unit cell dimensions for thorium nitrato complexes are listed in Table 46. Full structural data are available for several of the compounds. The complexes of the type $M''Th(NO_3)_6 \cdot 8H_2O$ contain the hexanitrato group $[Th(NO_3)_6]^{2-}$ and the hydrated cations $[M''(H_2O)_6]^{2+}$ linked by weak hydrogen bonding and are better formulated as $[M''(H_2O)_6]$-$[Th(NO_3)_6] \cdot 2H_2O$. Thus, in the magnesium complex, each thorium atom is bonded to twelve oxygen atoms (Th–O, average = 263 pm) at the corners of an irregular icosahedron (**Fig. 30**). A schematic representation of the structure is shown in **Fig. 31** [16]; selected bond lengths are listed in Table 47, p. 113.

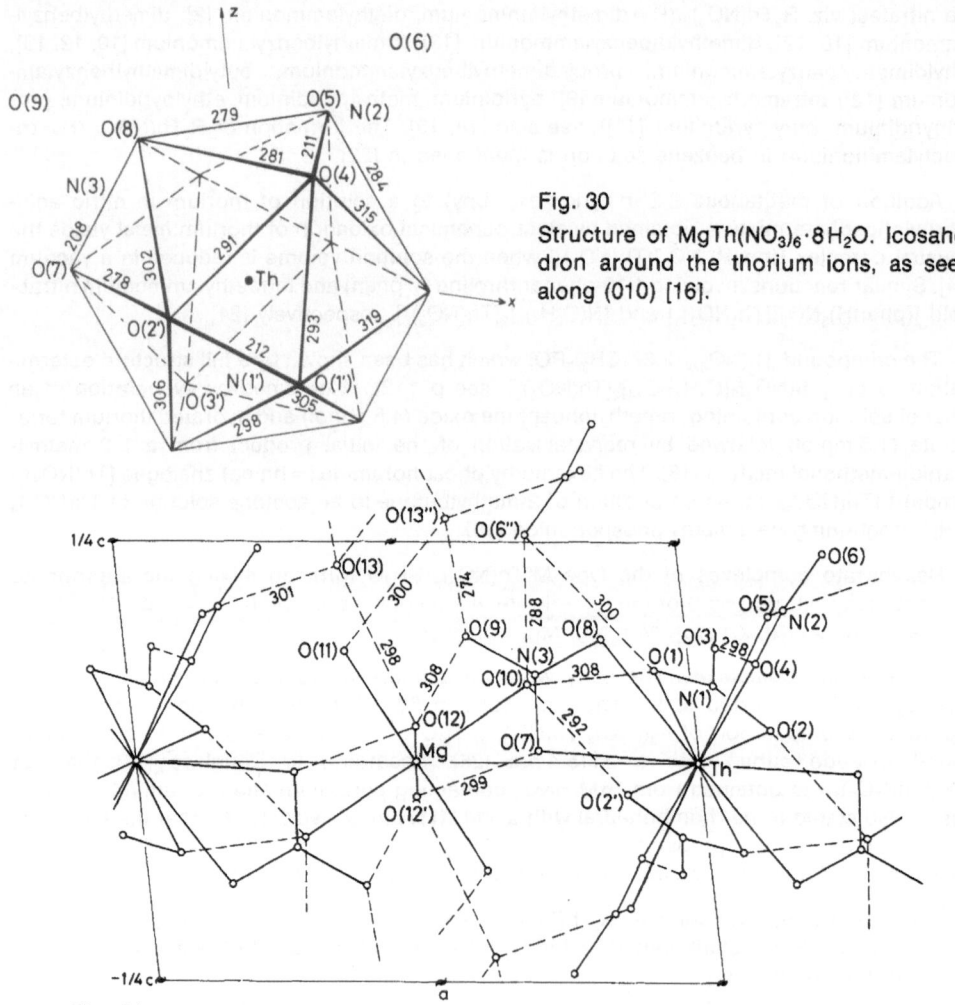

Fig. 30

Structure of $MgTh(NO_3)_6 \cdot 8H_2O$. Icosahedron around the thorium ions, as seen along (010) [16].

Fig. 31

Schematic representation of the structure of $MgTh(NO_3)_6 \cdot 8H_2O$ in the (010) projection. Nitrate ions are shown by heavy lines; Th–O and Mg–H$_2$O bonds are traced by thin lines. Dashed lines connect atoms of different polyhedra and isolated water molecules; only distances shorter than 310 pm are taken into account [16].

References for 4.7 see p. 123

Table 46

Crystallographic Data for Thorium Nitrato Complexes.

compound	symmetry	space group	lattice parameters in pm (or °) a (α)	b (β)	c (γ)	z	D(exp)	D(calc) in g/cm³	Ref.
$Rb_2Th(NO_3)_6$	monoclinic	$P2_1/c$ (No. 14)	834	689 β=121.2°	1531	2		3.42	[4]
$MgTh(NO_3)_6 \cdot 8H_2O$	monoclinic	$P2_1/c$ (No. 14)	908	875 β=97.0°	1361	4	2.41	2.39	[16]
$CoTh(NO_3)_6 \cdot 8H_2O$	monoclinic	$P2_1/c$ (No. 14)	908	878 β=97°	1362	4	2.49	2.49	[16]
$NiTh(NO_3)_6 \cdot 8H_2O$	monoclinic	$P2_1/c$ (No. 14)	908	876 β=97°	1363	4	2.52	2.52	[16]
$MnTh(NO_3)_6 \cdot 8H_2O$	monoclinic	$P2_1/c$ (No. 14)	908	875 β=97°	1361	4	2.50	2.49	[16]
$ZnTh(NO_3)_6 \cdot 8H_2O$	monoclinic	$P2_1/c$ (No. 14)	908	875 β=97°	1363	4	2.54	2.51	[16]
$[Th(NO_3)_3(tmpo)_4]_2[Th(NO_3)_6]$	monoclinic	Pn (No. 7)	952.3(1)	1450.7(2) β=95.322(13)°	2730.2(5)	2	1.93	1.93	[18]
$[Th(NO_3)_3(hmpa)_4]_2[Th(NO_3)_6]$	triclinic	$P\bar{1}$ (No. 2)	2476(5) α=73.4(5)°	2077(5) β=104.8(5)°	1288(5) γ=110.8(5)°				[19]
$[P(C_6H_5)_4][Th(NO_3)_5(tmpo)_2]$	orthorhombic	Pnma (No. 62)	2462.6(6)	1292.2(2)	1277.7(2)	4	1.75	1.74	[18]
$[(bpyH)_3NO_3][Th(NO_3)_6]$	triclinic	$P\bar{1}$ (No. 2)	930.0(2) α=108.79(2)°	1416.1(3) β=95.77(2)°	1703.7(4) γ=102.88(2)°	2	—	1.857	[20]
$K_3H_4[Th(NO_3)_{11}]$	rhombohedral	R3c (No. 161)	1111(1) α=90.5(1)°				2.440	2.505	[17]

References for 4.7 see p. 123

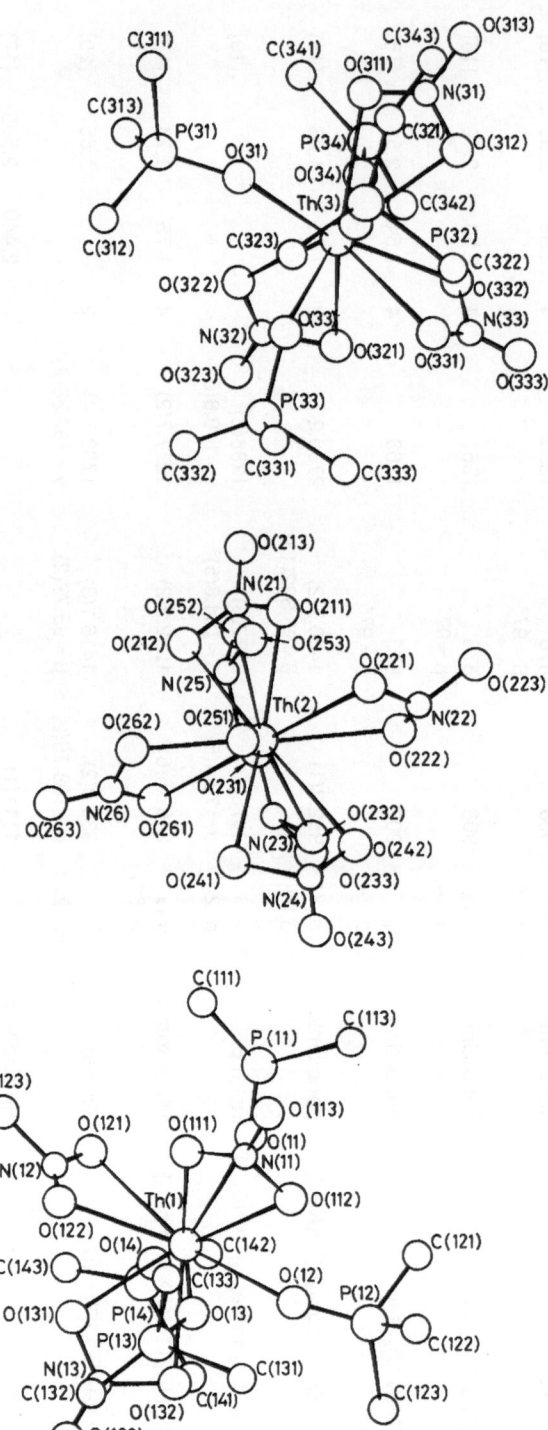

Fig. 32
The individual ions of
[Th(NO₃)₃(L)₄]₂[Th(NO₃)₆]
(L = trimethylphosphine oxide,
(CH₃)₃PO) showing the atomic
numbering scheme, viewed
along b [18].

References for 4.7 see p. 123

Table 47

Selected Bond Lengths (in pm) in MgTh(NO$_3$)$_6$·8H$_2$O (estimated accuracy ±5 pm) [16] (cf. Fig. 30, p. 110).

Th–O(1)	272		O(1)···O(7)	319
Th–O(2)	280		O(1)···O(8)	305
Th–O(4)	262		O(2)···O(4')	291
Th–O(5)	250		O(2)···O(5)	306
Th–O(7)	258		O(2)···O(7')	278
Th–O(8)	257		O(2)···O(8')	302
O(1)···O(2)	212	edges of	O(4)···O(7')	315
O(4)···O(5)	211	three NO$_3$	O(4)···O(8)	281
O(7)···O(8)	208	triangles	O(5)···O(7')	284
O(1)···O(4')	293		O(5)···O(8)	279
O(1)···O(5)	298			

Within the octahedron around magnesium

Mg–O(10)	212	O(10)···O(12)	291
Mg–O(11)	209	O(10)···O(12')	314
Mg–O(12)	216	O(11)···O(12)	300
O(10)···O(11)	298	O(11)···O(12')	300
O(10)···O(11')	296		

More recently the trimethylphosphine oxide and hexamethylphosphoramide complexes Th(NO$_3$)$_4$·2.67 L (L = ligand) have been shown to be ionic with the formulation [Th(NO$_3$)$_3$(L)$_4$]$_2$-[Th(NO$_3$)$_6$]. The structures are shown in **Fig. 32** and **Fig. 33**, respectively [18, 19].

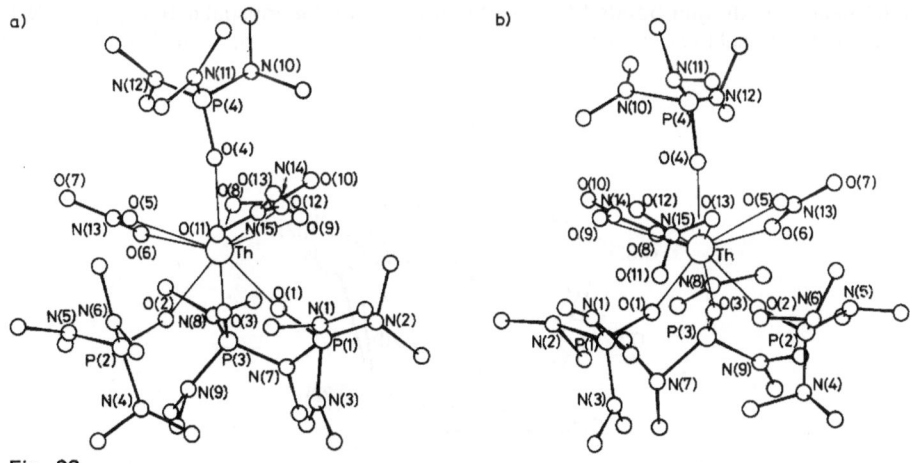

Fig. 33

The structures and atomic nomenclature of the cations A and B, and anions C and D in [Th(NO$_3$)$_3$(L)$_4$]$_2$[Th(NO$_3$)$_6$] (L = hexamethylphosphoramide, hmpa) [19].

The Th–O distances for the hexanitrato-groups range from 246 to 294 pm (average, 261 pm) in $[Th(NO_3)_3(C_3H_9PO)_4]_2[Th(NO_3)_6]$ and from 224 to 253 pm for the trimethylphosphine oxide ligands [18]. Selected bond lengths for $[Th(NO_3)_3(hmpa)_4]_2[Th(NO_3)_6]$ are listed in Table 48; more extensive data are available in the original papers together with details of bond angles.

Table 48

Selected Bond Lengths (in pm) in $[Th(NO_3)_3(hmpa)_4]_2[Th(NO_3)_6]$ (hmpa = hexamethylphosphoramide) [19] (cf. Fig. 33, p. 113).

	cation A	cation B		anion C	anion D
Th–O(1)	239(3)	232(3)	Th–O(1)	253(2)	254(3)
Th–O(2)	242(3)	231(2)	Th–O(11)	254(2)	246(3)
Th–O(3)	231(3)	236(4)	Th–O(2)	257(2)	256(3)
Th–O(4)	232(3)	230(3)	Th–O(21)	256(3)	264(2)
Th–O(5)	262(2)	252(5)	Th–O(4)	264(2)	251(2)
Th–O(6)	257(2)	237(6)	Th–O(41)	261(2)	252(3)
Th–O(8)	265(2)	239(4)	Th–O(5)	261(2)	258(3)
Th–O(9)	250(2)	292(6)	Th–O(51)	264(2)	258(3)
Th–O(11)	245(2)	—	Th–O(7)	267(3)	257(2)
Th–O(12)	259(2)	275(5)	Th–O(71)	262(3)	258(3)
Th–O(13)	—	264(3)	Th–O(8)	255(2)	257(3)
			Th–O(81)	260(2)	257(2)

The complex $[(bpyH)_3NO_3][Th(NO_3)_6]$ (bpy = 2,2'-bipyridine) contains $[Th(NO_3)_6]^{2-}$ anions, in which the arrangement of the twelve bonded oxygen atoms around the thorium atom is described as a nonsymmetric icosahedron with approximate C_2 symmetry (**Fig. 34**), and $[bpyH]^+$ cations hydrogen bonded to a nitrate ion to form the unusual cation $[(bpyH)_3NO_3]^{2+}$ [20]. The unit cell packing is shown in **Fig. 35** and selected bond lengths are listed in Table 49.

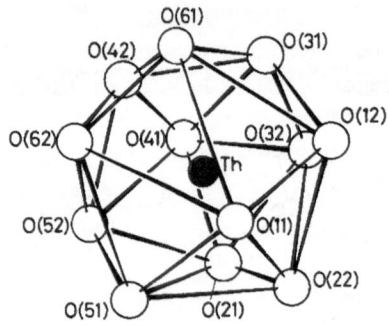

Fig. 34

The ThO_{12} polyhedron in $[(bpyH)_3NO_3][Th(NO_3)_6]$ showing approximate C_2 symmetry [20].

References for 4.7 see p. 123

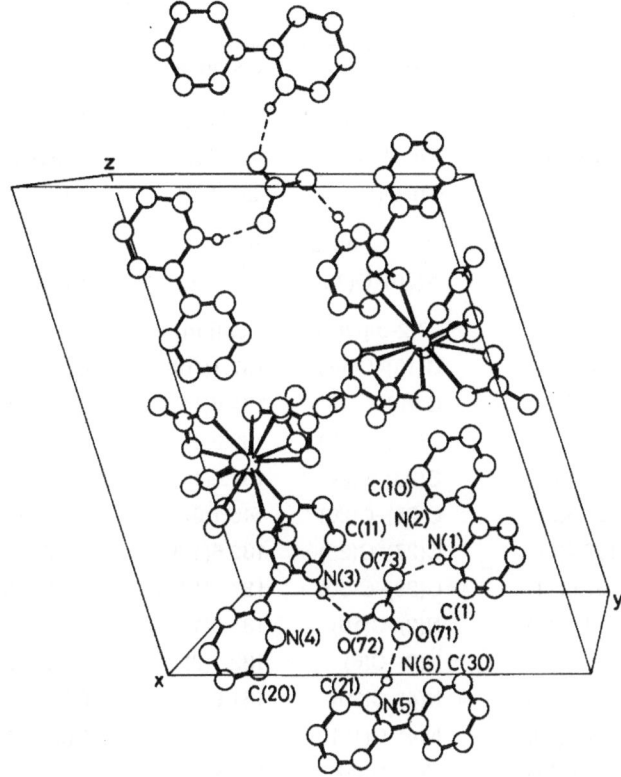

Fig. 35

Unit cell packing of [(bpyH)₃NO₃][Th(NO₃)₆]. Hydrogen bonds are shown by dotted lines. The symmetry-equivalent positions for bipyridinium cations with labels N(1)/N(2) are 1−x, 1−y, 1−z; with N(3)/N(4) x, y, z; and N(5)/N(6) x, y, z−1 [20].

Table 49

Interatomic Distances (in pm) with Estimated Standard Deviations for [(bpyH)₃NO₃][Th(NO₃)₆] [20].

atoms	distance	atoms	distance	atoms	distance
		[Th(NO₃)₆]²⁻			
Th–O(11)	256.8(3)	N(30)–O(32)	127.3(7)	Th–O(62)	257.3(4)
Th–O(21)	256.2(4)	N(30)–O(33)	119.5(8)	N(20)–O(21)	126.8(5)
Th–O(31)	257.2(4)	N(50)–O(51)	127.5(6)	N(02)–O(22)	127.9(7)
Th–O(41)	256.0(3)	N(50)–O(52)	127.4(6)	N(20)–O(23)	119.1(8)
Th–O(51)	255.1(4)	N(50)–O(53)	120.7(8)	N(40)–O(41)	127.0(6)
Th–O(61)	254.0(4)	Th–O(12)	256.6(4)	N(40)–O(42)	128.4(6)
N(10)–O(11)	127.5(8)	Th–O(22)	256.0(4)	N(40)–O(43)	120.7(5)

                References for 4.7 see p. 123   8*

Table 49 (continued)

atoms	distance	atoms	distance	atoms	distance
N(10)–O(12)	127.2(7)	Th–O(32)	258.5(4)	N(60)–O(61)	128.4(7)
N(10)–N(13)	121.4(6)	Th–O(42)	260.2(3)	N(60)–O(62)	127.2(8)
N(30)–O(31)	127.4(7)	Th–O(52)	257.3(4)	N(61)–O(63)	120.0(7)
		nitrate ion			
N(70)–O(71)	123.2(8)	N(70)–O(73)	125.2(6)	N(70)–O(72)	123.2(9)
		2,2-bipyridinium cation			
N(1)–C(1)	133.8(9)	C(16)–C(17)	137.1(10)	C(2)–C(3)	137.6(9)
N(1)–C(5)	135.5(7)	C(18)–C(19)	134.6(11)	C(4)–C(5)	138.9(10)
N(3)–C(11)	133.5(7)	C(21)–C(22)	138.3(8)	C(6)–C(7)	137.1(8)
N(3)–C(15)	136.1(8)	C(23)–C(24)	137.9(11)	C(8)–C(9)	137.8(13)
N(5)–C(21)	133.3(9)	C(25)–C(26)	146.1(9)	C(11)–C(12)	138.2(9)
N(5)–C(25)	136.4(8)	C(27)–C(28)	139.5(12)	C(13)–C(14)	138.3(8)
C(1)–C(2)	137.6(11)	C(29)–C(30)	137.2(12)	C(15)–C(16)	147.8(7)
C(3)–C(4)	138.5(10)	N(2)–C(10)	134.0(10)	C(17)–C(18)	139.5(10)
C(5)–C(6)	146.4(8)	N(2)–C(6)	132.7(9)	C(19)–C(20)	137.1(11)
C(7)–C(8)	136.1(12)	N(4)–C(20)	131.6(8)	C(22)–C(23)	136.7(10)
C(9)–C(10)	137.5(10)	N(4)–C(16)	133.0(8)	C(24)–C(25)	137.9(8)
C(12)–C(13)	137.6(11)	N(6)–C(26)	134.4(7)	C(26)–C(27)	136.6(10)
C(14)–C(15)	136.9(9)	N(6)–C(30)	134.1(11)	C(28)–C(29)	137.0(10)
		hydrogen bonding			
H(N1)–O(73)	187.6	N(3)–O(72)	283.8(9)	N(1)–O(73)	277.8(9)
H(N5)–O(71)	190.0	H(N3)–O(72)	193.7	N(5)–O(71)	281.3(9)

The thorium atom is also twelve co-ordinate in the complex $[P(C_6H_5)_4][Th(NO_3)_5(L)_2]$ (L = trimethylphosphine oxide) as shown in **Fig. 36** [18]. The Th–O distances for the nitrate groups range from 252 to 275 pm and those for the ligand groups are 234 and 242 pm, respectively (see Table 50).

Table 50
Bond Lengths (in pm) in $[P(C_6H_5)_4][Th(NO_3)_5(C_3H_9PO)_2]$ [18] (cf. Fig. 36).

Th–O(2)	242.1(12)	P(3)–C(3)	175.8(33)	P(1)–C(21)	180.2(18)
Th–O(3)	233.9(13)	P(3)–C(4)	181.8(25)	(1)–C(31)	181.4(17)
Th–O(11A)	252.4(15)	N(1)–O(11A)	135.5(19)	C(11)–C(12)	142.5(17)
Th–O(12A)	261.1(18)	N(1)–O(12A)	124.9(21)	C(11)–C(16)	139.2(17)
Th–O(11B)	274.3(18)	N(1)–O(11B)	119.7(21)	C(12)–C(13)	140.1(19)

References for 4.7 see p. 123

Table 50 (continued)

Th–O(12B)	275.3(18)	N(1)–O(12B)	127.6(21)	C(13)–C(14)	138.9(19)
Th–O(21A)	259.6(26)	N(1)–O(13)	121.9(16)	C(14)–C(15)	138.3(20)
Th–O(22A)	259.0(17)	N(2)–O(21A)	126.7(28)	C(15)–C(16)	140.1(19)
Th–O(21B)	270.5(25)	N(2)–O(22A)	132.1(20)	C(21)–C(22)	139.9(17)
Th–O(22B)	266.1(16)	N(2)–O(21B)	122.6(28)	C(22)–C(23)	140.4(21)
Th–O(31A)	261.7(15)	N(2)–O(22B)	124.6(18)	C(23)–C(24)	140.6(21)
Th–O(32A)	259.3(15)	N(2)–O(23)	123.5(15)	C(31)–C(32)	139.4(15)
P(2)–O(2)	148.5(13)	N(3)–O(31A)	125.9(20)	C(32)–C(33)	142.2(19)
P(2)–C(1)	179.3(18)	N(3)–O(32A)	126.5(20)	C(33)–C(34)	138.7(18)
P(2)–C(2)	183.0(27)	N(3)–O(33)	123.5(24)		
P(3)–O(3)	149.1(13)	P(1)–C(11)	179.4(11)		

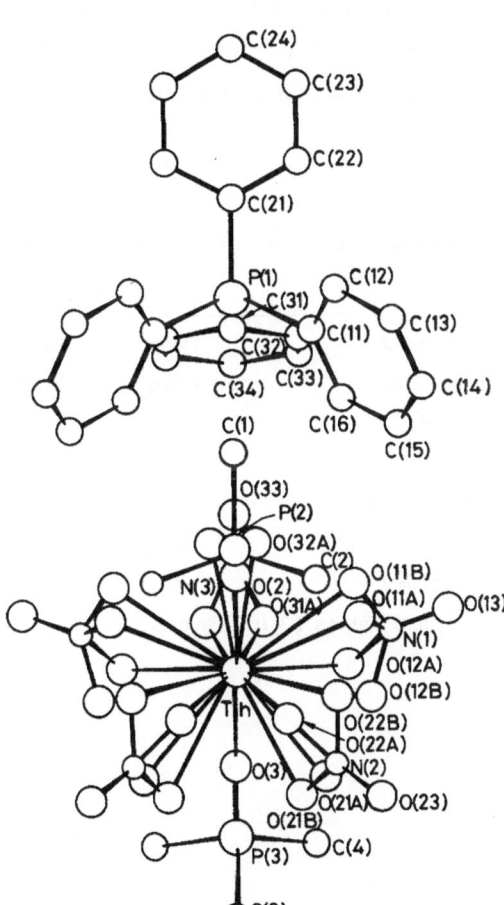

Fig. 36

The individual ions of $[P(C_6H_5)_4]$-$[Th(NO_3)_5(C_3H_9PO)_2]$ showing the atomic numbering (viewed along c with the mirror plane vertical) [18].

References for 4.7 see p. 123

Refractive indices for the various complexes are given in Tables 23 (p. 78), 51 and 52. Tables 23 and 51 also contain details of measured densities and molar volumes. Information on vibrational spectral studies of thorium nitrato complexes will be found in [6, 8, 18, 24, 26]. Selected infrared and Raman assignments are listed in Tables 53 and 54.

Table 51

Densities, Molar Volumes, and Refractive Indices for Some Thorium Nitrato Complexes [2] [*].

compound	mol. mass	D in g/cm^3 (25°C)	V$_m$ in cm^3/mol	refractive indices	
				n$_g$ (= n$_\gamma$)	n$_p$ (= n$_\alpha$)
$[CH_3NH_3]Th(NO_3)_5 \cdot 7H_2O$	700.04	2.429	288.2	1.549	1.504
$[CH_3NH_3]_2Th(NO_3)_6$	668.04	2.487	269.37	1.567	1.544
$[(CH_3)_2NH_2]Th(NO_3)_5 \cdot 8H_2O$	732.04	2.277	321.49		
$[C_2H_5NH_3]Th(NO_3)_5 \cdot 5H_2O$	678.04	2.424	279.72	1.561	1.514
$[(C_2H_5)_2NH_2]Th(NO_3)_5 \cdot 11H_2O$	814.04	2.088	407.02	1.542	1.522
$[(C_2H_5)_2NH_2]_2Th(NO_3)_6 \cdot 2H_2O$	788.04	2.055	383.47	1.543	1.522

[*] See also Tables 23 (p. 78) and 52.

Table 52

Some Optical Properties of Thorium Nitrato Complexes [3][*].

complex	principal refractive indices			2V
$(NH_4)_2Th(NO_3)_6$	1.588	1.599	1.613	84° (calc)
$Rb_2Th(NO_3)_6$	1.576	1.586	1.609	69°
$Cs_2Th(NO_3)_6$	1.579	1.586	1.613	34°
$MgTh(NO_3)_6 \cdot 8H_2O$	1.509	1.525	1.526	34°
$ZnTh(NO_3)_6 \cdot 8H_2O$	1.519	1.537	1.538	11.5°
$CoTh(NO_3)_6 \cdot 8H_2O$	1.521	1.540	1.540	9.5°
$NiTh(NO_3)_6 \cdot 8H_2O$	1.527	1.547	1.548	16.0°

[*] See also Tables 23 (p. 78) and 51.

References for 4.7 see p. 123

Table 53

Infrared Spectral Data for Thorium Nitrato Complexes (s = strong, vs = very strong, m = medium, w = weak).

compound	ν_1	ν_4	ν_2	ν_6	ν_3	ν_5	Ref.
$NaTh(NO_3)_5 \cdot 8.5 H_2O$	~1520vs*)	~1300vs	1036s	808s	746s	720s	[26]
$KTh(NO_3)_5 \cdot 6 H_2O$	~1550vs*)	~1300vs	1038, 1026	808s	748s	723s	[26]
$Rb_2Th(NO_3)_6$	~1530vs*)	~1300vs	1035s	813s	749s	728s	[26]
$Cs_2Th(NO_3)_6$	~1530vs*)	~1290vs	1037s	839s / 817s	750s / 728s	713m	[26]
$[(bpyH)_3NO_3][Th(NO_3)_6]$	1502s	1295s	1020m	805m	735s	708m	[24]
$[(phenH)_3NO_3][Th(NO_3)_6]$	1505s	1275s	1020s	818m	736m	728w	[24]
$MgTh(NO_3)_6 \cdot 8 H_2O$	~1520vs*)	1313vs / 1293vs / 1280vs	1037s	813m	750s	727s	[26]
$MnTh(NO_3)_6 \cdot 8 H_2O$	~1525vs*)	1300vs	1033s	818m	753m	723m	[26]
$CoTh(NO_3)_6 \cdot 8 H_2O$	~1530vs*)	1300vs	1038s	812m	751s	728s	[26]
$NiTh(NO_3)_6 \cdot 8 H_2O$	~1520vs*)	1300vs	1038s	818m	756s	730s	[26]
$K_3H_3Th(NO_3)_{10} \cdot 4 H_2O$	~1565vs*) / 1530vs*)	1330vs / 1302vs / 1290vs	1040s	838s / 818s / 810s	752s	730s	[26]
$[Th(NO_3)_3(C_3H_9PO)_4]_2[Th(NO_3)_6]$	1495w / 1515w	1300vw	1037s		763w / 750w	710w	[18]
$[P(C_6H_5)_4][Th(NO_3)_5(C_3H_9PO)_2]$	1520w	—	1038s / 1029m		756w	708w	[18]

*) Complex band.

References for 4.7 see p. 123

Table 54

Raman Data for Thorium Nitrato Complexes [26].

compound	ν_1	ν_4	ν_2	ν_6	ν_3	ν_5
$Rb_2Th(NO_3)_6$	1562(6)*) 1549(4) 1525(4)	1292(1)	1035(10)	808(1)	750(8)	710(3)
$Cs_2Th(NO_3)_6$	1560(6) 1515(5)	1335(1)	1040(10)	850(1) 818(1)	760(2) 726(6)	710(3)
$MgTh(NO_3)_6 \cdot 8\,H_2O$	1545(4) 1520(3) 1498(3)	1330(1)	1035(10)	827(2)	770(3) 753(4)	718(2)
$K_3H_3Th(NO_3)_{10} \cdot 4\,H_2O$	1546(3) 1495(4)	1330(1)	1057(6) 1032(10)	825(1)	712(2) 757(4)	724(1)

*) Relative intensities in parentheses.

4.7.3 Chemical Properties

The conversion of $(NO_2)_2Th(NO_3)_6$ to $Cs_2Th(NO_3)_6$ and $(NO)_2Th(NO_3)_6$ is described earlier (p. 109). The thermal decomposition of $(NO_2)_2Th(NO_3)_6$ and $(NO)_2Th(NO_3)_6$, to yield $Th(NO_3)_4$, is covered in Section 4.6.1 (p. 70) and the decomposition of $(NO_2)_2Th(NO_3)_6$ to give $ThO(NO_3)_2$ is described in Section 4.6.6, p. 99.

According to [1] on being heated in air $NaTh(NO_3)_5 \cdot 8.5\,H_2O$, and complexes of the type $M''Th(NO_3)_6 \cdot 8\,H_2O$ (M'' = Mg, Mn, Ni, Co) do not yield stable lower hydrates or the corresponding anhydrous complexes. $KTh(NO_3)_5 \cdot 6\,H_2O$ yields a pentahydrate in the temperature range 70 to 130°C and a phase of overall composition $KThO_{1.5}(NO_3)_2$ at 400°C; neither product was isolated and characterized [1]. $K_3(H_3O)_3Th(NO_3)_{10} \cdot H_2O$ is apparently converted to $K_3Th(NO_3)_7$ at ca. 130°C; this decomposes at about 400°C [1]. $Rb_2Th(NO_3)_6$ and $Cs_2Th(NO_3)_6$ are also reported to decompose at 400°C, yielding the appropriate alkali metal nitrate and thorium dioxide [1]. A more recent publication [7] indicates that $Cs_2Th(NO_3)_6$ decomposes over the temperature range 200 to 330°C; the enthalpy for the overall decomposition reaction is 225 kJ/mol. The activation energy decreases with progression of the reaction, levelling off to a value of ca. 190 kJ/mol above 50% decomposition.

With the exception of $Rb_2Th(NO_3)_6$, $Cs_2Th(NO_3)_6$ and $K_3Th(NO_3)_7$ all the complexes mentioned in the preceding paragraph are very hygroscopic; all the compounds dissolve readily in water and are soluble in nitric acid [1]. Solubility data for $Cs_2Th(NO_3)_6$ are illustrated in **Fig. 37** [13]. $NaTh(NO_3)_5 \cdot 8.5\,H_2O$ and the alkaline earth hydrated complexes are reported to dissolve in ethanol [1].

Complexes of the types $RTh(NO_3)_5 \cdot xH_2O$ (R = $CH_3NH_3^+$, x = 7; $(CH_3)_2NH_2^+$, x = 8; $C_2H_5NH_3^+$, x = 5; $(C_2H_5)_2NH_2^+$, x = 11) and $R_2Th(NO_3)_6 \cdot xH_2O$ (R = $CH_3NH_3^+$, x = 0; $(C_2H_5)_2NH_2^+$, x = 2) are readily soluble in water, methanol and ethanol; they are insoluble in toluene and other

References for 4.7 see p. 123

(unidentified) non-polar solvents [2]. Solubility data have not been recorded for these complexes but data are available for other hexanitrato complexes. Thus, the solubilities (in nitric acid) of complexes containing substituted ammonium and substituted pyridinium cations have been reported by Nikol'skii et al. [10 to 13] (see also [14, 15]). The results are listed in Tables 55 and 56 (p. 122); those for the dimethylbenzylammonium and trimethylbenzylammonium complexes are also illustrated in Fig. 37 and data for the dimethyldibenzylammonium complex are shown in **Fig. 38**.

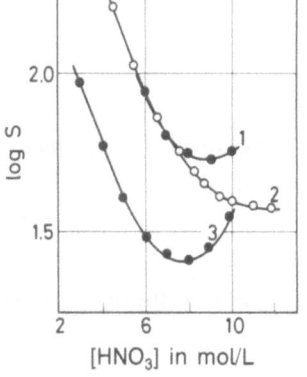

Fig. 37

Solubility S of:
1) $[DMBA]_2Th(NO_3)_6$; 2) $Cs_2Th(NO_3)_6$; 3) $[TMBA]_2Th(NO_3)_6$ in nitric acid [13].
DMBA = dimethylbenzylammonium;   TMBA = trimethylbenzylammonium.

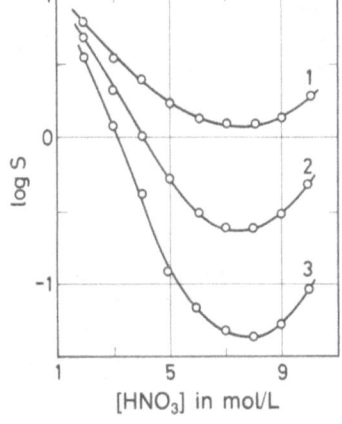

Fig. 38

Dependence of the solubility S of $[DMDBA]_2Th(NO_3)_6$ on HNO_3 concentration. $[DMDBA]NO_3$ concentration in solution: 1) 0; 2) 0.005; 3) 0.01M [13].
DMDBA = dimethyldibenzylammonium.

Table 55

Solubility of Alkyldimethylbenzylammonium Hexanitratothorates(IV) in Nitric Acid Solutions [12].

R in $R_2Th(NO_3)_6$	HNO_3 conc. in mol/L	Th^{4+} contents in mmol/L at the following temperature		
		25°C	40°C	50°C
	6.0	87.1	—	—
dimethylbenzylammonium	7.0	64.2	96.1	130.0
	8.0	56.6	92.3	124.1
	9.0	54.3	87.9	128.4

                    References for 4.7 see p. 123

Table 55 (continued)

R in $R_2Th(NO_3)_6$	HNO_3 conc. in mol/L	Th^{4+} contents in mmol/L at the following temperature		
		25°C	40°C	50°C
trimethylbenzylammonium	6.0	30.5	—	—
	7.0	27.4	40.4	52.6
	8.0	26.1	41.3	55.5
	9.0	28.2	46.1	64.2
ethyldimethylbenzylammonium	6.0	25.5	—	—
	7.0	21.7	33.8	45.9
	8.0	20.4	34.1	46.4
	9.0	22.6	37.8	56.8
propyldimethylbenzylammonium	6.0	11.1	—	—
	7.0	9.41	14.2	19.1
	8.0	9.38	15.1	20.8
	9.0	10.5	17.6	25.1
butyldimethylbenzylammonium	6.0	7.84	—	—
	7.0	7.02	11.5	16.2
	8.0	6.94	12.2	17.5
	9.0	7.80	13.8	21.7

Table 56

Solubility of Alkylpyridiniumhexanitratothorates(IV) [11].

R in $R_2Th(NO_3)_6$	HNO_3 in mol/L	Th content in saturated solution in mmol/L	R in $R_2Th(NO_3)_6$	HNO_3 in mol/L	Th content in saturated solution in mmol/L
pyridinium	8.0	793	propylpyridinium	6.0	105
	9.0	792		6.8	92.4
	10.0	813		8.0	87.9
				9.0	92.6
methylpyridinium	7.0	384			
	8.0	345	butylpyridinium	6.0	40.9
				6.8	35.6
ethylpyridinium	6.0	199		8.0	33.7
	6.8	184		9.0	30.5
	8.0	177			
	9.0	196			

References for 4.7:

[1] Molodkin, A. K., Belyakova, Z. V., Ivanova, O. M. (Zh. Neorgan. Khim. **16** [1971] 1582/9; Russ. J. Inorg. Chem. **16** [1971] 835/9).

[2] Molodkin, A. K., Ivanova, O. M., Belyakova, Z. V. (Zh. Neorgan. Khim. **16** [1971] 1461/2; Russ. J. Inorg. Chem. **16** [1971] 774/6).

[3] Staritzky, E., Truitt, A. L. (in: Seaborg, G. T., Katz, J. J., The Actinide Elements, McGraw Hill, New York 1954, pp. 797/838).

[4] Walker, D. I., Cromer, D. T., Staritzky, E. (Anal. Chem. **28** [1956] 1635/6).

[5] Val'tsev, V. K., Artamonova, S. M., Didora, N. F., Kravchenko, L. Kh. (Izv. Sibirsk. Otd. Akad. Nauk SSSR **4** [1961] 38/42; C.A. **1961** 207480).

[6] Alvey, P. J., Bagnall, K. W., Brown, D. (J. Chem. Soc. Dalton Trans. **1973** 2326/30).

[7] du Preez, J. G. H., van Vuuren, C. P. J., McGill, W. J. (J. Coord. Chem. **5** [1975/76] 231/5).

[8] Schmeisser, M., Koehler, G. (Angew. Chem. **77** [1965] 456).

[9] Ryan, J. L. (J. Phys. Chem. **64** [1960] 1375/85).

[10] Nikol'skii, B. P., Posvol'skii, M. V., Markov, G. S. (Radiokhimiya **8** [1966] 114/5; Soviet Radiochem. **8** [1966] 105/6).

[11] Nikol'skii, B. P., Posvol'skii, M. V., Markov, G. S. (Radiokhimiya **9** [1967] 713/4; Soviet Radiochem. **9** [1967] 671/3).

[12] Nikol'skii, B. P., Posvol'skii, M. V., Markov, G. S. (Radiokhimiya **9** [1967] 46/51; Soviet Radiochem. **9** [1967] 44/9).

[13] Nikol'skii, B. P., Markov, G. S., Posvol'skii, M. V. (Radiokhimiya **12** [1970] 272/8; Soviet Radiochem. **12** [1970] 246/51).

[14] Nikol'skii, B. P., Posvol'skii, M. V., Markov, G. S. (Radiokhimiya **16** [1974] 711/5; Soviet Radiochem. **16** [1974] 699/703).

[15] Nikol'skii, B. P., Posvol'skii, M. V., Markov, G. S. (Proc. Moscow Symp. Chem. Transuranium Elem., Moscow 1972 [1976], pp. 239/42).

[16] Šćavničar, S., Prodič, B. (Acta Cryst. **18** [1965] 698/702).

[17] Douglass, R. M. (Anal. Chem. **29** [1957] 165/6).

[18] Alcock, N. W., Esperas, S., Bagnall, K. W., Wang, H-Y. (J. Chem. Soc. Dalton Trans. **1978** 638/46).

[19] English, R. P., du Preez, J. G. H., Massimbeni, L. R., van Vuuren, C. P. J. (S. African J. Chem. **32** [1979] 119/25).

[20] Khan, M. A., Kumar, N., Tuck, D. G. (Can. J. Chem. **62** [1984] 850/5).

[21] van Uitert, L. G., Gradkiewicz, W. H. (Mater. Res. Bull. **6** [1971] 283/91).

[22] Dem'yanova, T. A., Lipovskii, A. A., Spivakov, B. Ya. (Radiokhimiya **14** [1972] 898/900; Soviet Radiochem. **14** [1972] 933/5).

[23] Ferraro, J. R., Katzin, L. I., Gibson, G. (J. Am. Chem. Soc. **77** [1955] 327/8).

[24] Kumar, N., Tuck, D. G. (Can. J. Chem. **62** [1984] 1701/4).

[25] Volkov, Yu. F., Kapshukov, I. I., Vasil'ev, V. Ya. (Tezisy Dokl. 1st Vses. Konf. Khim. Urana, Moscow 1974, p. 26).

[26] Petrov, K. I., Molodkin, A. K., Saralidze, O. D., Belyakova, Z. V. (Zh. Neorgan. Khim. **12** [1967] 2974/82; Russ. J. Inorg. Chem. **12** [1967] 1573/8).

Table of Conversion Factors

Following the notation in Landolt-Börnstein [7], values which have been fixed by convention are indicated by a bold-face last digit. The conversion factor between calorie and Joule that is given here is based on the thermochemical calorie, cal_{th}, and is defined as 4.1840 J/cal. However, for the conversion of the "Internationale Tafelkalorie", cal_{IT}, into Joule, the factor 4.1868 J/cal is to be used [1, p. 147]. For the conversion factor for the British thermal unit, the Steam Table Btu, BTU_{ST}, is used [1, p. 95].

Force	N	dyn	kp
1 N (Newton)	1	10^5	0.1019716
1 dyn	10^{-5}	1	1.019716×10^{-6}
1 kp	9.80665	9.80665×10^5	1

Pressure	Pa	bar	kp/m^2	at	atm	Torr	lb/in^2
1 Pa (Pascal) $=1N/m^2$	1	10^{-5}	1.019716×10^{-1}	1.019716×10^{-5}	0.986923×10^{-5}	0.750062×10^{-2}	145.0378×10^{-6}
1 bar $=10^6$ dyn/cm^2	10^5	1	10.19716×10^3	1.019716	0.986923	750.062	14.50378
1 kp/m^2=1mm H$_2$O	9.80665	0.980665×10^{-4}	1	10^{-4}	0.967841×10^{-4}	0.735559×10^{-1}	1.422335×10^{-3}
1 at $=1$kp/cm^2	0.980665×10^5	0.980665	10^4	1	0.967841	735.559	14.22335
1 atm $=760$ Torr	1.01325×10^5	1.01325	1.033227×10^4	1.033227	1	760	14.69595
1 Torr $=1$mm Hg	133.3224	1.333224×10^{-3}	13.59510	1.359510×10^{-3}	1.315789×10^{-3}	1	19.33678×10^{-3}
1 lb/in^2=1 psi	6.89476×10^3	68.9476×10^{-3}	703.069	70.3069×10^{-3}	68.0460×10^{-3}	51.7149	1

Work, Energy, Heat	J	kWh	kcal	Btu	MeV
1 J (Joule) = 1 Ws = 1 Nm = 10^7 erg	1	2.778×10^{-7}	2.39006×10^{-4}	9.4781×10^{-4}	6.242×10^{12}
1 kWh	3.6×10^6	1	860.4	3412.14	2.247×10^{19}
1 kcal	4184.0	1.1622×10^{-3}	1	3.96566	2.6117×10^{16}
1 Btu (British thermal unit)	1055.06	2.93071×10^{-4}	0.25164	1	6.5858×10^{15}
1 MeV	1.602×10^{-13}	4.450×10^{-20}	3.8289×10^{-17}	1.51840×10^{-16}	1

$1 \, eV \triangleq 23.0578$ kcal/mol = 96.473 kJ/mol

Power	kW	PS	kp·m/s	kcal/s
1 kW = 10^{10} erg/s	1	1.35962	101.972	0.239006
1 PS	0.73550	1	75	0.17579
1 kp·m/s	9.80665×10^{-3}	0.01333	1	2.34384×10^{-3}
1 kcal/s	4.1840	5.6886	426.650	1

References:

[1] A. Sacklowski, Die neuen SI-Einheiten, Goldmann, München 1979. (Conversion tables in an appendix.)
[2] International Union of Pure and Applied Chemistry, Manual of Symbols and Terminology for Physicochemical Quantities and Units, Pergamon, London 1979; Pure Appl. Chem. **51** [1979] 1/41.
[3] The International System of Units (SI), National Bureau of Standards Spec. Publ. 330 [1972].
[4] H. Ebert, Physikalisches Taschenbuch, 5th Ed., Vieweg, Wiesbaden 1976.
[5] Kraftwerk Union Information, Technical and Economic Data on Power Engineering, Mülheim/Ruhr 1978.
[6] E. Padelt, H. Laporte, Einheiten und Größenarten der Naturwissenschaften, 3rd Ed., VEB Fachbuchverlag, Leipzig 1976.
[7] Landolt-Börnstein, 6th Ed., Vol. II, Pt. 1, 1971, pp. 1/14.
[8] ISO Standards Handbook 2, Units of Measurement, 2nd Ed., Geneva 1982.

Key to the Gmelin System
of Elements and Compounds

System Number	Symbol	Element
1		Noble Gases
2	H	Hydrogen
3	O	Oxygen
4	N	Nitrogen
5	F	Fluorine
6	**Cl**	**Chlorine**
7	Br	Bromine
8	I	Iodine
8a	At	Astatine
9	S	Sulfur
10	Se	Selenium
11	Te	Tellurium
12	Po	Polonium
13	B	Boron
14	C	Carbon
15	Si	Silicon
16	P	Phosphorus
17	As	Arsenic
18	Sb	Antimony
19	Bi	Bismuth
20	Li	Lithium
21	Na	Sodium
22	K	Potassium
23	NH_4	Ammonium
24	Rb	Rubidium
25	Cs	Caesium
25a	Fr	Francium
26	Be	Beryllium
27	Mg	Magnesium
28	Ca	Calcium
29	Sr	Strontium
30	Ba	Barium
31	Ra	Radium
32	**Zn**	**Zinc**
33	Cd	Cadmium
34	Hg	Mercury
35	Al	Aluminium
36	Ga	Gallium

System Number	Symbol	Element
37	In	Indium
38	Tl	Thallium
39	Sc, Y	Rare Earth
	La—Lu	Elements
40	Ac	Actinium
41	Ti	Titanium
42	Zr	Zirconium
43	Hf	Hafnium
44	Th	Thorium
45	Ge	Germanium
46	Sn	Tin
47	Pb	Lead
48	V	Vanadium
49	Nb	Niobium
50	Ta	Tantalum
51	Pa	Protactinium
52	**Cr**	**Chromium**
53	Mo	Molybdenum
54	W	Tungsten
55	U	Uranium
56	Mn	Manganese
57	Ni	Nickel
58	Co	Cobalt
59	Fe	Iron
60	Cu	Copper
61	Ag	Silver
62	Au	Gold
63	Ru	Ruthenium
64	Rh	Rhodium
65	Pd	Palladium
66	Os	Osmium
67	Ir	Iridium
68	Pt	Platinum
69	Tc	Technetium[1]
70	Re	Rhenium
71	Np, Pu . . .	Transuranium Elements

HCl
$CrCl_2$
$ZnCrO_4$
$ZnCl_2$

Material presented under each Gmelin System Number includes all information concerning the element(s) listed for that number plus the compounds with elements of lower System Number.

For example, zinc (System Number 32) as well as all zinc compounds with elements numbered from 1 to 31 are classified under number 32.

[1] A Gmelin volume titled "Masurium" was published with this System Number in 1941.

A Periodic Table of the Elements with the Gmelin System Numbers is given on the Inside Front Cover